THE SANDSTONE POWDER MAGAZINE AT THE BENICIA ARSENAL

BENICIA'S LITTLE-KNOWN GEM

The Sandstone Powder Magazine at the Benicia Arsenal

Benicia's Little-known Gem

H. Allan Gandy

//
Alive Book Publishing

The Sandstone Powder Magazine
at the Benicia Arsenal
Copyright © 2021 by H. Allan Gandy

All rights reserved. No part of this book may be reproduced or transmitted in any form or by any means without written permission from the publisher and author.

Additional copies may be ordered from the publisher for educational, business, promotional or premium use. For information, contact ALIVE Book Publishing at: alivebookpublishing.com, or call (925) 837-7303.

Book Design by Alex P. Johnson
Cover Photo: Library of Congress, HABS no. CA-1948-3
Back Cover Photo by the Author
About the Author photo by Susan Gandy

ISBN 13
978-1-63132-131-3

Library of Congress Control Number: 2021906610
Library of Congress Cataloging-in-Publication Data is available upon request.

First Edition

Published in the United States of America by ALIVE Book Publishing and ALIVE Publishing Group, imprints of Advanced Publishing LLC
3200 A Danville Blvd., Suite 204, Alamo, California 94507
alivebookpublishing.com

PRINTED IN THE UNITED STATES OF AMERICA

10 9 8 7 6 5 4 3 2 1

Acknowledgements

The Benicia Historical Museum was essential in providing research and archive materials. The museum's digital archives contain many photos used in this book. The Benicia Arsenal ledger detail activities between 1856 and 1859 providing insight to the many men who built the powder magazine.

Many thanks to Dr. James E. Lessenger, research historian, for donating his time for editorial suggestions, tips on writing style, organization of chapters and providing historical details. I am grateful to my friends and colleagues at the museum for their help and support: executive director Elizabeth d'Huart, archives director Beverly Phelan, board member Larry Lauber, board member Elizabeth Murphy, and research associate Bob Kwasnicka.

Special thanks to James Kilby Merrill, great, great grandson of Caleb Strong Merrill, for providing early family history.

To my wife Susan, the love of my life.

Contents

Introduction..11

Chapter One: The Benicia Historical Museum Complex............................13

Chapter Two: The Design of the Stone Powder Magazines.......................15

Chapter Three: The Benicia Arsenal is Established...................................19

Chapter Four: The Powder Magazine Agreement of 1855.........................25

Chapter Five: Caleb S. Merrill, Builder of the Powder Magazines............29

Chapter Six: The Powder Magazine Building 2..33

Chapter Seven: The Powder Magazine Building 10...................................45

Chapter Eight: The Sandstone...51

Chapter Nine: The Eagle and Cannon Sculpture on Magazine 10............57

Chapter Ten: The Building of Stone Powder Magazine 10........................67

Chapter Eleven: The Powder Magazine 10 Exterior..................................81

Chapter Twelve: The Powder Magazine 10 Interior...................................93

Chapter Thirteen: The Fire of 1922...109

Chapter Fourteen: Saving the Powder Magazine Building 10................115

Chapter Fifteen: The Powder Magazines Vaulted Ceiling History.........119

Chapter Sixteen: Magazine 10 Today...123

Appendix A – List of Men Who Built Powder Magazine 10....................127

Appendix B – List of Inscriptions on the Walls..131

Introduction

The two stone powder magazines at the Benicia Arsenal, also referred to as building 2 and building 10, were built in 1855 and 1857, respectively. Each was designed to hold three thousand barrels of powder. They are noted for their handsomely carved sandstone block exterior walls four feet in depth, and the beautiful interior vaulted structures of stone.

This book is focused on the powder magazine building 10 built in 1857 which is currently part of the Benicia Historical Museum complex in Benicia, California. It contains many photographs showing the intricacies of the unique features on the magazine while telling the story of its design, construction and the men who built it.

Chapter One

The Benicia Historical Museum Complex

The land for the Benicia Historical Museum complex was acquired by the City of Benicia following the Arsenal closure in 1964. The following map (Figure 1) shows the buildings and sandstone quarries on a topographical map of the Military Reservation from 1868.

In 1868, this area was grass covered rolling hills with no trees. Several small creeks in the area were normally dry during summer and fall. There were two sandstone outcroppings that were utilized to quarry the sandstone blocks to build the warehouses and powder magazines.

The Benicia Historical Museum complex consists of the following buildings:

- Building 7, Storehouse
- Building 8, Engine House
- Building 9, Ammunition Shop
- Building 10, Powder Magazine

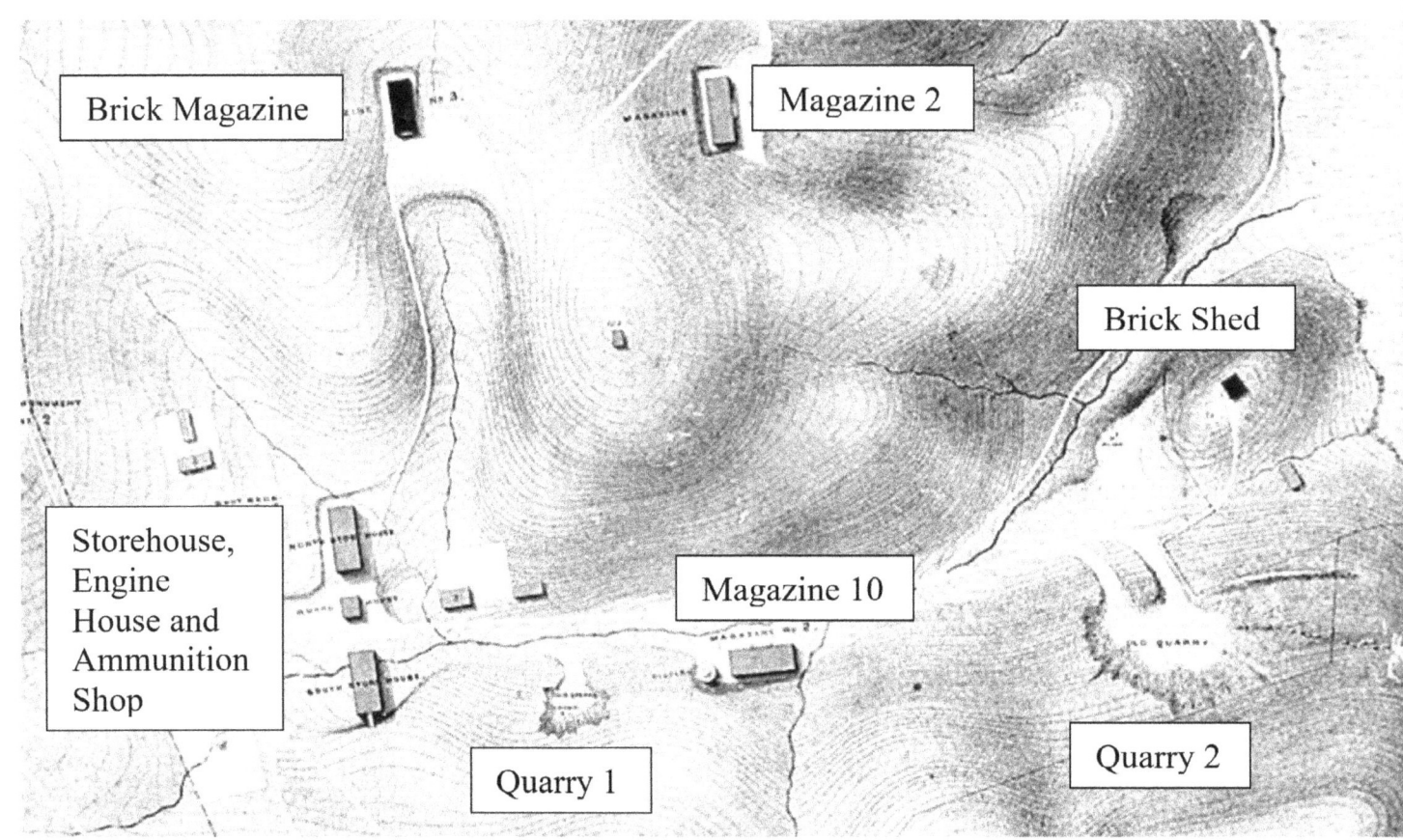

Figure 1: Storehouse, Engine House, Ammunition Shop, Powder Magazine, and Sandstone Quarries, 1868. Source: Benicia Historical Museum.

Chapter Two

The Design of the Stone Powder Magazines

The powder magazines were designed to be safe and nearly impenetrable. The interior vaulted ceiling design was selected so that if an explosion was to occur inside, the four foot thick walls would project the blast upward, thus limiting lateral damage. If the ceiling was breached, the keystones would collapse and the detonation would theoretically implode.[1] The magazines were placed in a location to make them less visible and accessible and to ensure that in the event of an explosion there were no other structures nearby.[2] Magazine 10 is located approximately 600 feet east of the Camel Barn Museum buildings, nestled between two low lying hills and a small creek that used to empty into Suisun Bay.

The powder magazines were originally designed in 1828 by the US Army at Watervliet Arsenal in New York. The Benicia Arsenal obtained the drawings from Commander Alfred Mordecai at Watervliet which were redrawn and used to build magazines 2 and 10.[3]

A construction drawing of Magazine 10 (Figure 2) shows a series of beams whose ends rest on the front and back walls, just clearing the crown of the vaults. The drawing only shows a portion of the front view and a cross section showing the interior elevation measurements, the roof support design and the space below the main floor (a network of sandstone blocks) designed to keep the floor and stored powder dry.

The following is a summary of the powder magazine's design and construction features:

- They utilized local sandstone for building
- The sandstone foundations and walls are four feet thick
- They are constructed with air space below the floor and utilize vents to keep them dry
- The ashlar style masonry means that no cement or bonding materials were used; the stones were cut to precisely fit together
- They were designed to direct any internal explosion upwards, thus minimizing lateral damage
- Two windows are at each end of the buildings with window doors inside and outside and four vents at mid-wall

- Decorative sandstone blocks, called quoins are at entrance and building corners
- A sandstone carved Eagle and Cannon is over the entrance to Magazine 10 dated 1857
- The interiors have beautiful arched stone ceilings, 9 feet, 4 inches at the highest point
- The center pillars have decorated carved capitals
- The interiors have wooden wainscoting walls
- The bolts and roof nails were made of copper alloy by Arsenal blacksmiths.

Drawings were prepared depicting the magazine design in 1976 as part of the Historic American Buildings Survey (HABS) and are shown below in Figures 3 and 4.

On each end wall of the building there are two square windows, measuring 32 by 32 inches with coverings, both inside and outside. The interior of the building consists of a large room, divided down the center by a row of six octagonal columns supporting a series of cross vaults.[4] The interior walls of the magazines have wooden wainscoting.

On the front and back of the magazine are two small rectangular holes connected with horizontal passages in the walls, which were used for ventilation (Figure 5). The design allows for air to enter the magazine but prevented objects from directly entering.

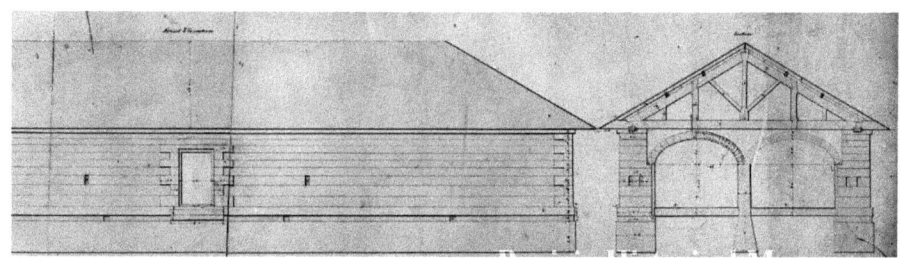

Figure 2: Drawing for the Magazine 10.
Source: Library of Congress, HABS no. CA-1948 Image 10.

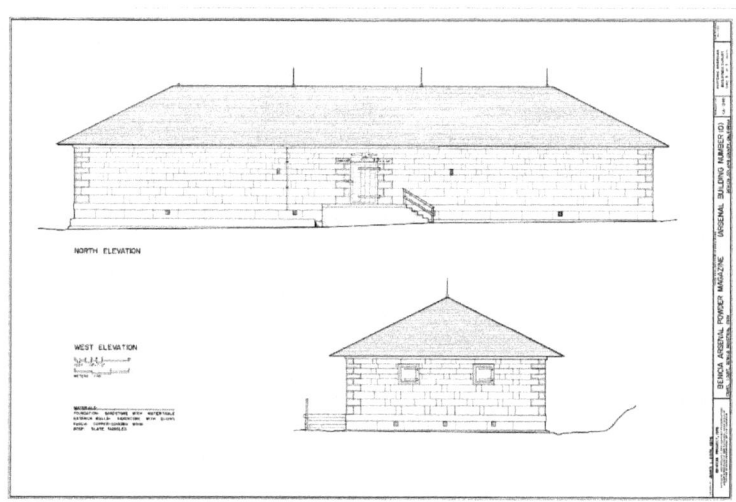

Figure 3: Drawing of the Magazine 10.
Source: Library of Congress, HABS Cal, 48-Beni Sheet 3.

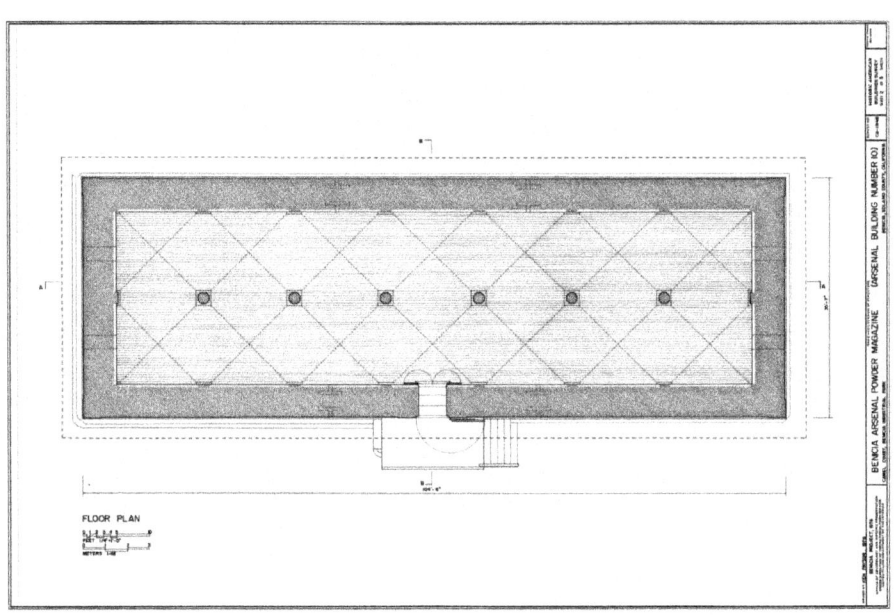

Figure 4: Magazine 10 floor plan with four foot thick walls.
Source: Library of Congress, HABS Cal, 48-Beni Sheet 2.

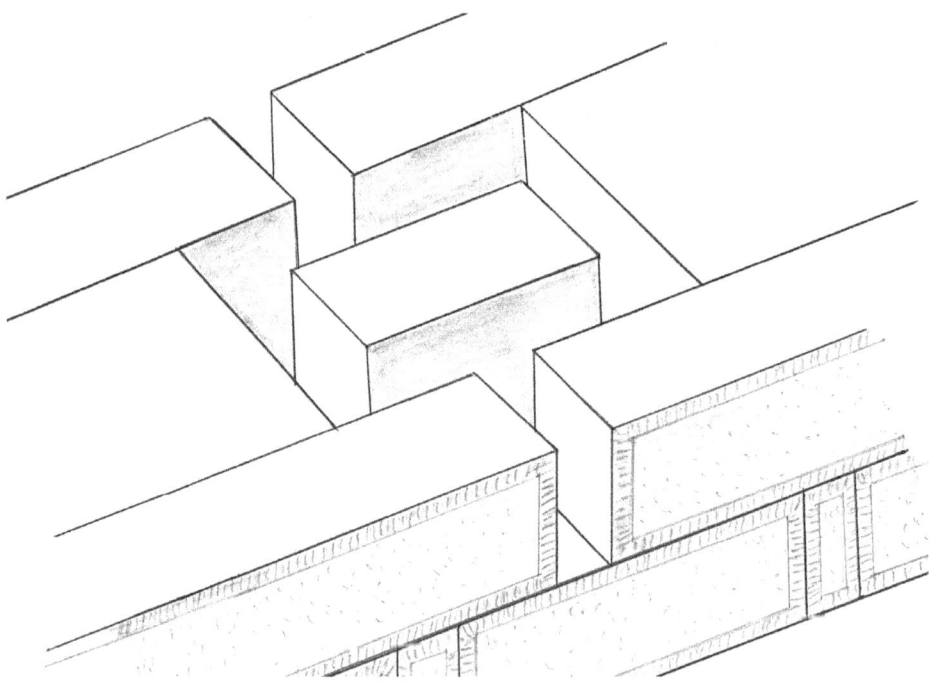

Figure 5: Vent passage through wall.
Source: Diagram by author.

Chapter Three

The Benicia Arsenal is Established

The story of the Benicia Arsenal begins in the days when military transportation was either by horse or by foot, and when military fire power was limited to cannon ball and black gunpowder.[5] In 1847, a 252-acre parcel of land adjoining the Benicia city limits on the east was acquired for a military reserve. The US Army chose Benicia (as opposed to San Francisco) because it offered several advantages. Having to store black gunpowder, Benicia offered a drier, less foggy climate for warehousing Army munitions. Its location was at a strategic point between the San Francisco and San Pablo Bay, and near the confluence of the Sacramento and San Joaquin rivers. Benicia was close enough to furnish ordnance to the San Francisco Bay Area, and its 30 mile inland location made it less vulnerable to attack. It could handle deepwater vessels and more easily supply the interior of the state via waterways and other transportation routes.[6] Importantly, it was free.

The first companies of the Army, C and G companies of the 2nd Infantry Regiment, occupied the military reservation on April 9, 1849 under command of Lt. Col. Silas Casey and set up camp to establish the Benicia Barracks.[7]

In 1851, a 27 year-old officer Brevet Captain Charles Pomeroy Stone from Fort Monroe, Virginia, was ordered to California by Brigadier General George Talcott, the Chief of Ordnance in Washington, DC, to establish an ordnance supply depot in the newly established military post in Benicia. His assignment was as follows:

> *Your duties in California will be, in general terms, the charge and supervision in duties of all ordnance property in the Pacific Division, and all business there pertaining to the Ordnance Department. On your arrival, you will report to the General commanding the division; informing him of your means both in men and materials, and consulting him as to the best position for the establishment of an Ordnance Depot in California.*
>
> *I deem it unnecessary to give you more detailed instructions, as your own judgement and professional information will readily supply them.*[8]

Captain Stone arrived in Benicia on August 15, 1851 aboard a small ship, the *Helen McGraw,* coming from New York and transporting muskets, rifles, sabres, swords, bayonets, spare parts, armored tools, shot, howitzers, cannon, substance supplies, and lumber.[9]

Stone discovered that the barracks which had been erected on the reservation by the Quartermaster's Department to be excellent for quartering his 21 enlisted men. He chose a site to the east of the Barracks for building the structures which would become the California Ordnance Depot.[10] The rolling hills of Benicia are covered with wild grass that is a beautiful green from the winter rains, but by the end of spring, dry to a golden color which become a wildfire danger by summer, a potentially hazardous situation to the military reservation that stored munitions.

A soldier stationed there in the 1850s said of the site:

> *In Benicia the climate is glorious. It seldom gets too warm. The ocean breeze makes it pleasant. The fogs so prevalent in San Francisco never comes this far. There is no rain during the summer months. There are but two seasons here, the dry and the rainy.*[11]

Soon after arrival in California, Captain Stone became aware of the deplorable state of ordnance supplies on the West Coast. In his letter to Lt. Colonel H. K. Craig, August 25, 1851, he reported:

> *From many official sources I learn that very considerable quantities of ordnance stores are scattered over this country and most of them in very bad condition for want of proper care and store rooms.*[12]

In need of a powder magazine, a temporary wooden structure, measuring 40 by 20 feet, was erected in 1851 before the winter rains. A three-man detail guarded the powder magazine 24 hours a day. Order 37 outlined the Rules and Regulations for the magazine:

- *Never to be opened except in the presence of our officer or some non-commissioned officer specifically designated by the commanding officer.*
- *Whoever enters the magazine must take off his shoes, or put on slippers over them and leave outside his cane, sword, sabre, or any other thing about him which can produce a spark.*

- *Whenever a magazine is opened, a guard must be stationed near with orders to allow no one to pass without permission. The sentry must be armed with a sabre or bayonet, never his fire-arm.*
- *Barrels of powder must never be rolled or dragged on the floor, but must always be lifted and carried in a handbarrow or on rods.*
- *When work is to be done in a magazine (moving barrels, etc.) the floor must be covered with a tarpaulin or be slightly sprinkled with water; all stones and metal must be carefully cleaned out, barrels must never be repaired, headeded or unheadeded in the magazine.*
- *Powder carried from the magazine in carts must be carefully packed on the bottom in straw wedged on a tarpaulin and carefully covered with a tarpaulin or cloth.*[13]

Captain Stone became increasingly aware of the necessity for a fireproof magazine. He wanted a secure magazine like the one at Watervliet Arsenal in New York. In late January 1852, he requested funds from the Chief of Ordnance. He reported:

> *Six to eight tons of powder are now stored in Benicia in the temporary magazine and the Commanding Officer at San Francisco is constantly urging of the removal here of the 970 barrels which are stored in an adobe building with a wooden roof, adjoining the gun carriage shed and stable.*[14]

While Stone's requests wound their way through the Army's bureaucracy, weapons began arriving to Benicia from up and down the West Coast. The powder issue was resolved temporarily by placing barrels aboard schooners in Suisun Bay for the dry (fire) season and then moving them ashore under what cover could be found once the fall and winter rains began.[15]

Local materials for construction (cement and lumber) were costly in California and bricks were generally of poor quality. Stone's observation was that the Military Reservation had a sufficient quantity of excellent sandstone that could be used to produce sandstone blocks, hand-hewn from the Arsenal hills with which to build. These structures would include a hospital, a guardhouse, five magazines, two shops (later to be known as the Camel Barns), a three story storehouse (later to be known as the Clock Tower), and a wharf.[16]

Frustrated with Washington's slow response to provide funds for building adequate facilities in Benicia, Stone took matters into his own hands in 1853 and used his own money. He initiated construction of a permanent ordnance wharf, two large two-story warehouses (Buildings 7 and 9) and a small engine house (Building 8).[17]

In 1854, the Benicia Arsenal had a complement of two officers, 48 enlisted men and a few civilian employees.[18] The permanent stone powder magazine had not yet been built. An inspection of the Arsenal conducted by a colonel sent by the Secretary of War found the lack of a fireproof structure disturbing. The temporary wooden structure was bulging under the weight of the stored powder and, despite the 24 hour guard, it was susceptible to fire, either from sabotage or more likely through carelessness.[19]

The construction timeline of the Arsenal buildings is shown in Table 1. The Arsenal buildings were assigned new building numbers in the early 1900s by the Army, often leading to confusion in earlier documents or references.

Table 1: Benicia Arsenal Buildings 1852-1859.

New Building No.	Old Building No.	Use	Construction Material	Date
-	3	Powder Magazine	Brick and wood	1852
1	-	Post Hospital	Sandstone	1856
2	1	Powder Magazine	Sandstone	1855
7	-	Warehouse	Sandstone	1853
8	-	Engine House	Sandstone	1855
9	-	Warehouse	Sandstone	1854
10	2	Powder Magazine	Sandstone	1857
29	-	Warehouse	Sandstone	1859

Because of the high cost of living and low pay, Army officers, particularly junior officers, were not discouraged from supplementing their incomes with private-sector work.[20] So Stone, in addition to fulfilling his professional duties at the Benicia Arsenal, engaged in railroad company work and gold bullion brokering. The former venture failed but Stone was successful in the gold and banking business. Stone was financially ruined when a senior partner mismanaged several loans and an office clerk absconded with a large sum of money.[21]

Stone resigned from the Army in November 1856 intending to dedicate all of his time and effort to supporting his family and trying to make good on his mounting debts. He never saw the completion of the second stone powder magazine. He was replaced by Captain Franklin D. Callender.[22]

In the following years, Stone was reportedly the first volunteer to enter the Union Army in January 1861. In this role, he secured the capital for the arrival of President-elect Abraham Lincoln and was personally responsible for security at the new president's inaugural.[23] During the Civil War he served as a brigadier general, noted for his involvement at the Battle of Ball's Bluff in October 1861. Held responsible for the Union defeat, Stone was arrested and imprisoned for almost six months, mostly for political reasons. He never received a trial, and after his release he would not hold a significant command during the war again. He later served again with distinction as a general in the Egyptian Army and is also known for his role in constructing the base of the Statue of Liberty.[24]

Figure 6: Brevet Captain Charles Pomeroy Stone, circa 1855. Source: Benicia Historical Museum.

Chapter Four
The Powder Magazine Agreement of 1855

Captain Stone hired the master mason, Caleb S. Merrill, to build the first stone powder magazine in 1855. Sydney N. Smith, Attorney-at-Law in San Francisco drew up an agreement between the Government of the United States and Caleb S. Merrill that was executed on October 30, 1855 (Figure 7).

The agreement outlined a detailed timeline of construction milestones and payments that were to be made by the Army upon completion. The text of this agreement is as follows:

> *This Agreement made and entered into on the Thirteenth day of October AD. Eighteen hundred Fifty five, between Brevit Captain Charles P. Stone, United States Ordnance Corps, Commanding the United States Arsenal at Benicia, California, for and in behalf of the Government of the United States, Party of the first part, and Caleb S. Merrill, Master Mason, Party of the second part, Witnesseth that for and in consideration of the sums hereinafter named, to be paid to him as hereinafter set forth, the said Party of the second part, agrees and binds himself, his heirs, executors and administrators, to furnish the stone for, and too build all the Stone work of a certain Powder Magazine to be erected at Benicia Arsenal, according to a plan approved in September 1855, by the Bvt. Captain Stone, commanding the forenamed Arsenal, and now deposited in the Arsenal Office at Benicia aforesaid, and according to the specifications prepared by the said Capt Stone, and annexed hereto, both which plans and specifications are hereby considered as forming part of this agreement, as fully as if the same were incorporated herein.*
>
> *And the said Party of the first part on behalf of the government of the United States, hereby agrees in consideration of the aforesaid undertakings on part of the said Party of the second part, to pay the said Party of the second part the sum of Nineteen thousand, nine hundred and thirty eight Dollars ($19,938) at the times and the sums following, that is to say:*
>
> *Eight hundred (800) Dollars when Five thousand (5000) feet of stone shall be hauled to the grounds and accepted.*

<u>Sixteen hundred Dollars</u> ($1600) when the Stones of four (4) Arches shall be cut.

<u>Five hundred Dollars</u> ($500) when One thousand (1000) feet of cutting of corners and jambs and piers and the water table and cornice stone be cut.

<u>Five hundred Dollars</u> ($500) when the Water table shall be finished.

<u>Eleven hundred Dollars</u> ($1100) when the foundation shall be completed.

<u>Sixteen hundred Dollars</u> ($1600) when the Stones for Arches shall be cut.

<u>Sixteen hundred Dollars</u> ($1600) when Fifteen thousand (15000) feet of stone have been delivered and accepted.

<u>Seven hundred Dollars</u> ($700) when the walls shall have been carried up four feet above the foundation.

<u>One thousand Dollars</u> ($1000) when the walls have been carried up to the springing lines of the arches and piers are up.

<u>Five thousand Dollars</u> ($5000) when the Arches shall be completed and the balance of <u>Five thousand five hundred</u> and <u>thirty eight</u> ($5,538) Dollars when the contract shall be fully completed and the work accepted.

<u>In witness whereof</u> the Parties of the first and second hereto have herein scribed their names and affixed their seals the day and year heretofore written.[25]

<u>Seals delivered in presence of</u>:
Evan Williams

Signatures and seals:
Charles P. Stone, Bvt Capt. Ord. Corps
C. S. Merrill

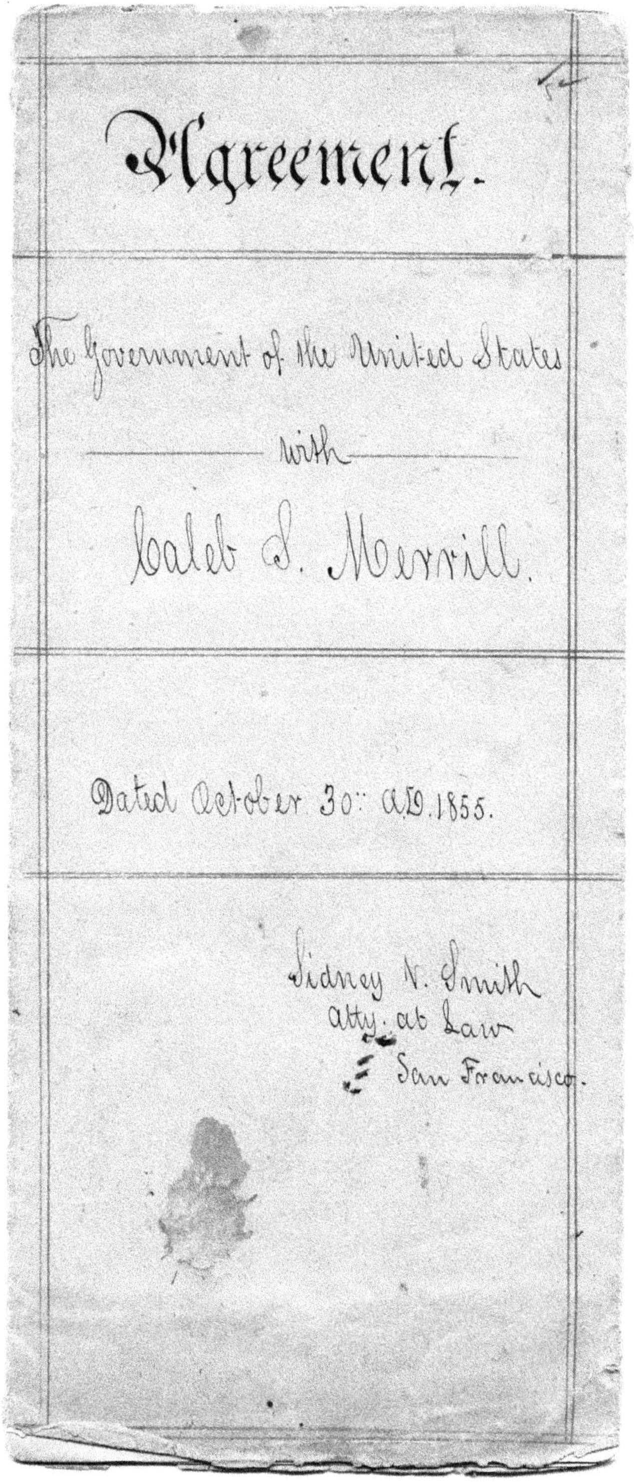

Agreement.

The Government of the United States

with

Caleb S. Merrill

Dated October 30th A.D. 1855.

Sidney V. Smith
Atty. at Law
San Francisco.

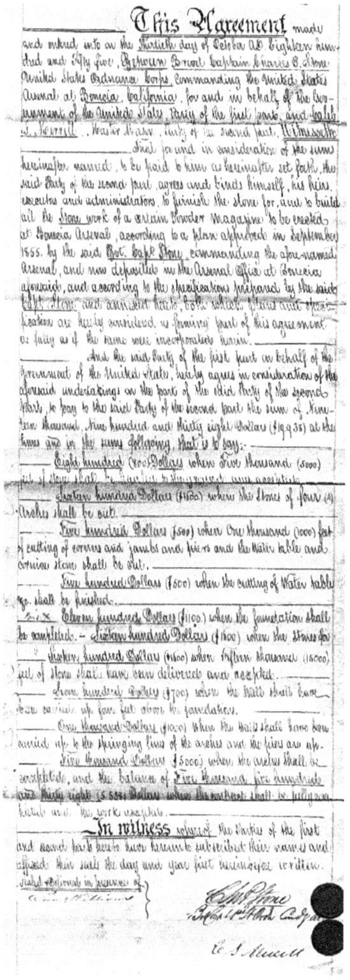

Figure 7: Agreement, October 30, 1855. Source: Benicia Historical Museum.

Chapter Five
Caleb S. Merrill, Builder of the Powder Magazines

Caleb Strong Merrill was the builder of the Powder Magazines 2 and 10. He was born September 25, 1806, in Shelburne Falls, Massachusetts. His father, Thaddeus, was a stone mason and Caleb took up the occupation in Shelburne.[26] In 1830, at the age of 24, he traveled to California on a "Boston hide drogher," a sailing ship used in the cowhide trade up and down the West Coast.[27] For the next three years he worked as a stone mason in repairing several of the California missions.[28] Then he returned home to Massachusetts and married Maria Childs in June of 1834. Caleb Strong, Jr. was born in September 1834. In 1836 the family moved to Ohio and their second child, Martha Eliza, was born in March 1837. The Merrill family moved to Monmouth, Illinois, in 1838, where they had three more children, Maria, Sabra and Asa.

Merrill was an architect and builder and had a contractor business with his partner, Lysander Woodworth, in Warren County, Illinois in 1839. They obtained a contract to build a new jail in Monmouth and completed it at the end of 1840.[29] Merrill may have worked on brick buildings being constructed in Monmouth, including the church building that became part of the founding of Monmouth College.

His wife Maria died at the young age of 34 in 1845, the year after her youngest son, Asa, was born. After his wife died, Merrill brought his young family of five back to Shelbourne Falls and they lived near his father, Thaddeus, age 70. Merrill continued to work as a mason in the area.

In 1852, Merrill and his 17-year-old son, Caleb Jr., traveled to California across the plains with a large party that endured many hardships. Many of the animals belonging to the party gave out and the emigrants were left with but one team, and had to walk the rest of the long journey. They reached California in the fall, tired and footsore.[30]

Merrill partnered with Pierre "Peter" Larseneur to build the wooden Benicia Barracks[31] in 1852 and in November 1853 Merrill obtained a contract to build a kitchen.[32] In 1854, they built the Old St. Mary's Cathedral in San Francisco (Figure 8) which still stands today on the corner of California Street and Grant Avenue. The stone for the cathedral came from Patrick Dillon in Benicia. Dillon came to Benicia in 1851, ran the Pioneer Stone Business in San Francisco and supplied sandstone from his quarry at his farm in the area now known as Dillon Point.[33]

In 1855, Merrill entered into a contract with the Army to build the first stone powder magazine. He oversaw the quarry work to excavate and remove sandstone blocks and the construction of the magazine. After completion of the magazine, Merrill and Larseneur were hired by the Army as a stone masons to build the second stone magazine in 1857. His son, Caleb Jr. was also employed by the Army as a stone mason and quarryman. Larseneur had two brothers, Charles and Louis, who were also hired to work on the magazine. Larseneur's daughter, Jenevier "Jennie" married Caleb's son in 1859 in Benicia.[34] Merrill continued to work at the Benicia Arsenal along with his son. Then in 1859, he was the master mason in building a stone store house, now known as the "Clock Tower."

Peter Larseneur lived in San Francisco in the 1860s where he did marble work for Andrew Paltenghi and Company.[35] He later joined a partnership with the Pioneer Steam Marble Works Company.[36] He died April 2, 1867 at the Geyser Hotel in Sonoma County at the age of 53.

After his work for the Army, Merrill followed the building and construction business in California with his son. He worked in the San Francisco Bay area in the early 1860s then moved to the San Joaquin Valley in 1866, running a freighting business. In the Spring of 1867, he was hired by representatives of the Pacific Development Company to haul an oil rig down the West Side of the San Joaquin Valley. The oil rig consisted of a hand-power pole, auger drilling rig, casing and lumber, and three wagons, two of which were driven by Merrill and his son. Later, they were hired by the Buena Vista Petroleum Company to reconstruct the road from Reward in the San Joaquin Valley across the mountains to San Luis Obispo.[37]

Caleb's youngest son Asa enlisted in the Union Army, Company H of the Massachusetts 10th infantry, on June 21, 1861. He was wounded and died in the Civil War of Battle of Fair Oaks, Virginia on May 31, 1862, at the age of 18.[38]

In the late 1870s, Merrill moved to Albion, Nebraska to be with his daughter, Mariah age 41, then later moved to Grundy County, Missouri in 1883. Merrill died at his home on April 6 in Grundy at the age of 76. His daughter wrote to friends that "her father was very much in love with Monmouth" and it was probable at some future date they would bring his remains to Monmouth to rest by his wife.[39] He was buried alongside his wife in Monmouth, Illinois later that year.

Caleb Merrill Jr. remained in Benicia until June 1861. He bought property in Rockville and engaged in the wine business, but it was not to his liking. He sold out in 1862 and found work in Santa Clara as a bookkeeper for two years. He moved to Santa Barbara where he went into the sheep business and was very successful. In May 1865, he moved to San Luis Obispo county securing 6,000 acres for his ranch. In 1877, he sold his land and bought the Avenal Ranch in Kings

County where he started raising cattle. He was again very successful and bought and traded until the ranch was several thousand acres. He lived out his life in Kings County and passed away in 1898.[40]

Figure 8: Old St. Mary's Cathedral in San Francisco, 1869.
Source: OpenSFHistory.org.

Chapter Six
The Powder Magazine Building 2

Powder Magazine Building 2 (Figure 9) was constructed in 1855 at a cost of $30,012. It is located approximately 800 feet northeast of the Building 7. It was constructed of fine rusticated sandstone ashlar with walls over 4 feet thick and a wood floor. Ashlar is a masonry style of stone cutting in which rectangular blocks are precisely cut and polished on all faces adjacent to those of other stones, so they can be fitted together without a bonding material. The roof was originally wood trusses with corrugated iron panels. The outside dimension of the building are 104 feet long by 35.6 feet wide, with approximately 2,414 square feet on the inside. It was used to store ammunition, including Class D explosives (ammonium picrate) and black powder.[41] Building 2 is nearly identical in design to Building 10, except for the lack of elaborate stone sculptural carvings on the interior columns (Figures 10 and 11).[42]

The Magazine Building 2 has deteriorated over the years. The slate roof was damaged in 1922 as a result of a projectile from a nearby magazine that exploded. The slate was replaced with galvanized corrugated iron roof in 1923.

Following closure of the Arsenal, the magazine was used for the storage of furniture. In 1964, the former site became the Benicia Industrial Park under the ownership of the city of Benicia. In 1974, the property around Magazine 2 was transferred to Benicia Industries, Inc (currently Amports) as part of a land swap. The roof was again destroyed by fire in 1983 and the magazine still stands but is not currently in use nor are there any plans to restore it (Figure 12). The front entrance is sealed to prevent any internal vandalism (Figure 13 and 14).

The Magazine 2 exterior sandstone walls have weathered extensively, primarily due to its location: The entrance side faces east toward the Suisun Bay which is unprotected from the Carquinez Strait winds (Figure 15 and 16).

The interior of the Magazine 2 has suffered from neglect and vandalism. The wooden wall wainscoting is damaged and the floor has deteriorated in many areas (Figure 17). Despite the poor condition, Powder Magazine building 2 is listed in the United States Department of the Interior, National Register of Historic Places, dated November 7, 1976. It is described as:

Magazine (Building #2)

This sandstone block building was constructed in 1855 to serve as a storehouse for gunpowder. It is a one story building measuring 35.6' x 104' with stone foundation. The corrugated roof is hipped in form. There is 2400 square feet of usable space - ceiling is 9' 6" in height. Floors are made of wood. The door opening measures 4' 2" in width. The interior is similar to building #10. Presently it is used for storage. The present roof replaced the original slate roof. Location is on Fir Road near Patrol Road.

Magazine 2 is on private property owned by Amports, Inc. and is not accessible to the public. The magazine stands along the 680 Freeway and is visible near the 680/780 interchange (Figure 18).

Figure 9: Powder Magazine, Building 2, 1961.
Source: Benicia Historical Museum.

Figure 10: Magazine 2 (left) columns are square and lack stylized carving while Powder Magazine 10 (right) columns are octagonal and have decorative carvings. Source: Photograph by author.

Figure 11: Magazine 2 (top) keystones are square while Magazine 10 (bottom) keystones have a cross design. Source: Photograph by author.

Figure 12: Magazine 2 roof damage from 1983 fire.
Source: Photograph by author.

Figure 13: Magazine 2 sealed front entrance.
Source: Photograph by author.

Figure 14: Magazine 2 exterior window doors.
Source: Photograph by author.

Figure 15: Magazine 2 extensive weathering of east facing wall.
Source: Photograph by author.

Figure 16: Magazine 2, northeast corner of building showing weathering. Source: Photograph by author.

Figure 17: Magazine 2 interior.
Source: Photograph by author.

Figure 18: Magazine 2 visible from freeway 680/780 interchange.
Source: Photograph by author.

Chapter Seven
The Powder Magazine Building 10

Building 10 was constructed in 1857 at a cost of $35,262. It is nearly identical in design to Magazine 2 except additional interior sculptural work. It was constructed of fine rusticated sandstone ashlar with stone walls over 4 feet thick. The roof was built with wood trusses and slate. The interior has six carved stone octagonal columns, wood floors and wood wall coverings. The outside dimensions of the stone building are 104 feet long by 35.6 feet wide, with approximately 2,460 square feet of storage space. It could store up to 3,000 barrels of black powder.[43] The magazine historical photos are shown in Figures 19 through 24.

Electrical power was brought to the magazine in about 1908 providing much needed interior lighting. The magazine escaped damaged in a 1922 arsenal fire that destroyed many other structures.

The magazine was used for gunpowder storage until about 1939. The magazine was cleaned out revealing the majestic beauty of the interior (Figure 24). When the arsenal was closed in 1964, the magazine sat idle for many years. After the Benicia Arsenal was listed on the National Register of Historical Places in 1976, the City of Benicia designated the magazine as part of the Benicia Historical Museum complex. The Powder Magazine 10, Building 7 Storehouse, and Building 9 Ammunition Shop were restored and dedicated in 1985. The roof was restored in 1991 by the museum using composite shingles.

Figure 19: Magazine 10 view of north front and east side, 1961.
Source: Benicia Historical Museum.

Figure 20: Magazine 10 view of north front and west side, 1977.
Source: HABS, photo by Sirlin Studios, Sacramento, CA.

*Figure 21: Cistern on west side of Magazine 10, 1963
(the cistern was removed during freeway construction in 1964).
Source: Benicia Historical Museum.*

*Figure 22: Magazine 10 entrance and sign, 1975.
Source: Library of Congress, HABS CA-1948 Image 2.*

*Figure 23: Magazine 10 entrance with eagle and cannon.
Source: Library of Congress, HABS CA-1948 Image 8.*

*Figure 24: The beautiful groin vaulted ceiling of Magazine 10.
Source: Library of Congress, HABS no. CA-1948-3.*

Chapter Eight
The Sandstone

Prior to the 20th Century, stone was the predominant material used in building construction. It was not only the structural material, but also the exterior and interior finish, and often the flooring and roofing as well. The properties of desired building stone includes strength, hardness, workability, porosity, durability, and appearance. The strength of a stone depends on its structure, the size and hardness of its particles, and the manner in which those particles are interlocked or cemented together. The strength is generally characterized by the compressive strength. Workability refers to the ease with which a stone may be sawed, shaped, dressed, or carved.[44]

The sandstone used for construction of the Camel Barns and powder magazines was quarried from two local hillsides; one small quarry just 300 feet behind the Camel Barns and a larger quarry 1,300 feet away (Figure 1). The geology in this area is Cretaceous Period bedrock of the Great Valley Sequence, 66 to 100 million years old.[45]

Sandstone is a sedimentary rock comprising an aggregate of sand-sized (0.06 to 2.0-mm) fragments of minerals and rocks held together by a mineral cement (Figure 25). Sandstone forms when sand is buried under successive layers of sediment. During burial, the sand is compacted and a binding agent, which in this area is clay, is precipitated from ground water which moves through passageways between grains during formation to cement the sandstone.[46] These sedimentary rocks were deposited during the late Cretaceous Period in an ancient seaway that corresponds roughly to the outline of the modern Central Valley of California.[47] The sandstone of this region is uniform and shows evidence of derivation from areas composed dominantly of granitic rock.[48]

This sandstone is composed primarily of quartz and plagioclase feldspar grains which are held together with a clay matrix which makes up to 50 percent of the sandstone's volume. The sandstone draws its yellowish-brown color from the character of the mineral composition and clay binding agent. Although the sandstone has a small pore structure, the clay matrix is soluble by acids which are in rainwater and is the prime contributor to sandstone deterioration.[49]

This sandstone is relatively soft and sand grains can easily be liberated when abraded by a sharp object. But the ground on which the arsenal is built is hard and solid and the foundations of the buildings have been well laid. As a consequence, they have suffered little from earthquakes.[50]

Analysis of the sandstone[51] provided the following composition:

- 52% clay matrix
- 21% quartz
- 16% plagioclase feldspar
- 6 % chert lithics
- 3% metamorphic lithics
- 1% volcanic lithics
- 1% tourmaline.

The inherent properties of this sandstone make it ideal for building and stone work. These properties are:

- It is fine grained and uniform in consistency (good durability)
- It is compact and has good compressive strength (strong)
- It is relatively soft, easy for workability (stone carving)
- It has low water absorption properties and is able to withstand repeated freeze-thaw cycles (less prone to cracking)
- It is fireproof (ideal for a powder magazine).

This sandstone was perfect for carving ashlar style, where the surfaces of each stone (and adjacent stones) are worked to a smooth, close fitting finish. The face is finished with a fine "sparrow pecked" and margin finish (Figure 26).[52]

This sandstone, being formed in an ancient seaway, contains small fossils. Fossils are visible in the sandstone blocks on the south facing wall, since this wall is weathered more than the other walls (Figure 27).

The powder magazine, being in a sheltered location nestled between two hills, shows fairly limited deterioration of the sandstone. There is minor deterioration at the base and wall on the south side and at the corners of the building (Figure 28). The joints between the stones are in good condition.[53]

Figure 25: Closeup of the fine-grained sandstone from the Arsenal quarry.
Source: Photograph by author.

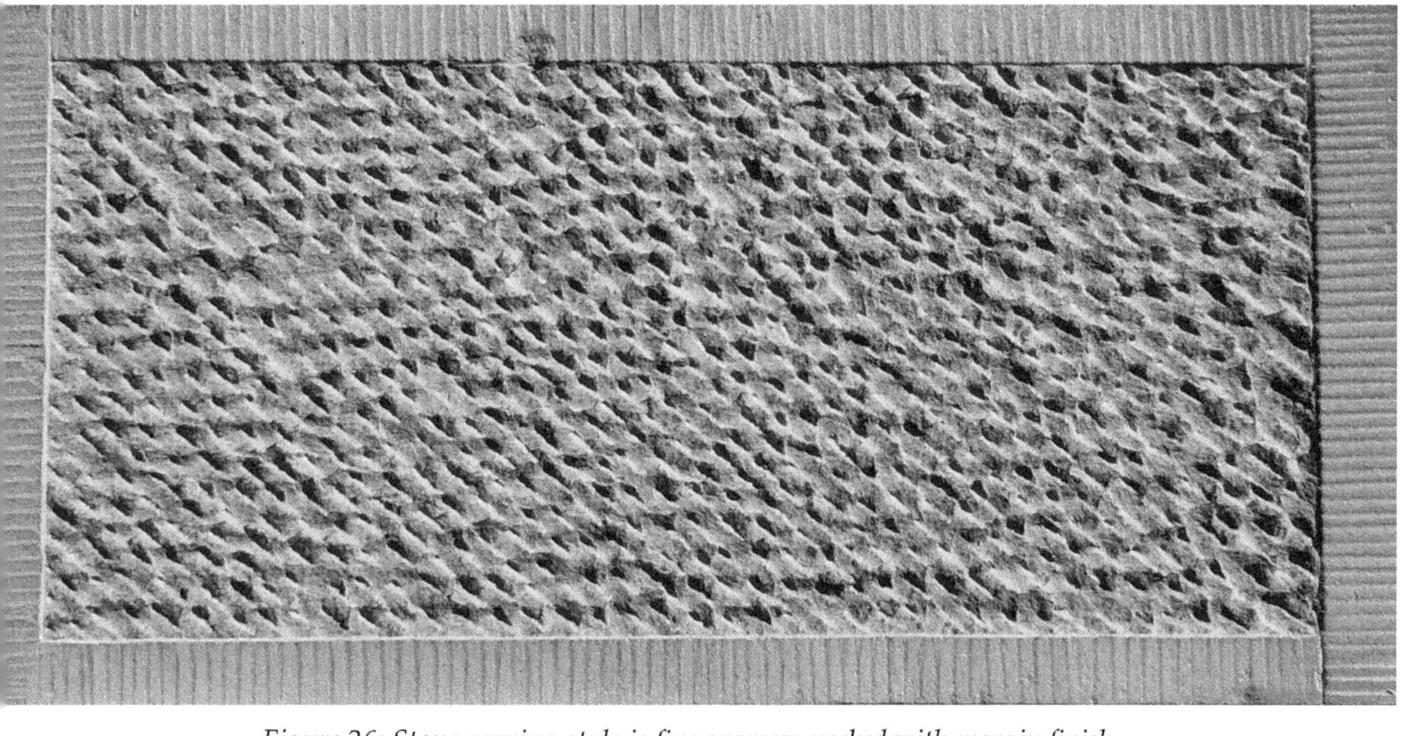

Figure 26: Stone carving style is fine sparrow pecked with margin finish.
Source: Photograph by author.

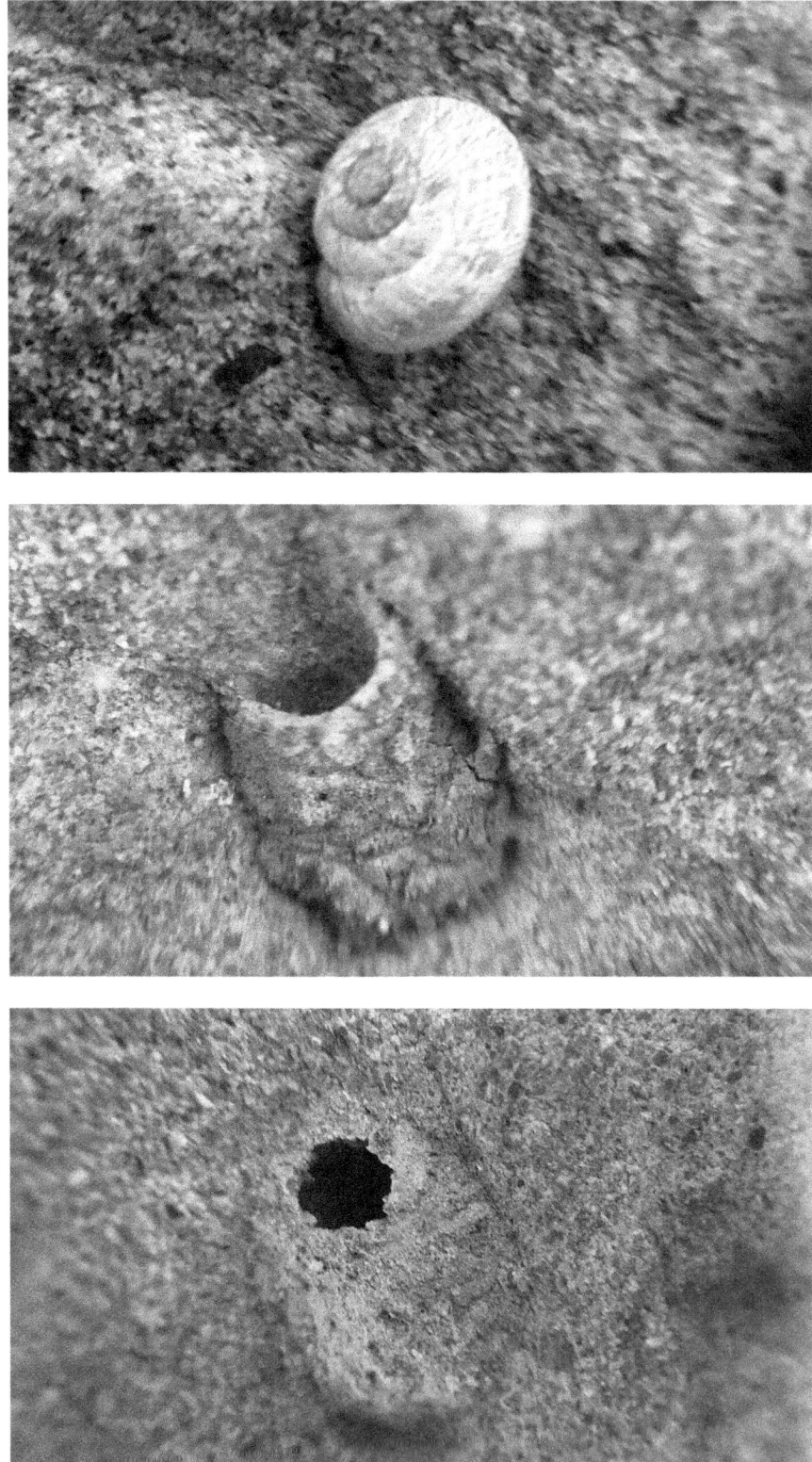

Figure 27: Sandstone fossils in south wall block:
Gastropod 1/8", (top), Baculite, 3/16" (center), Cirripedia 3/16" (bottom).
Source: Photograph by author.

*Figure 28: Sandstone deterioration at the base of Magazine 10 south side.
Source: Photograph by author.*

Chapter Nine

The Eagle and Cannon Sculpture on Magazine 10

One of the most artistic features of the Magazine 10 is the eagle and cannon sculpture above the entrance (Figure 29). To the left and right of the cannon is the statement "Erected A.D. 1857." In 1976, the Historic American Buildings Survey created a drawing of the sculpture to complement the registration of the Benicia Arsenal, which the powder magazine is one of the historic structures (Figure 30).

The eagle and cannon motif was carved from a single block measuring 10 feet 2 inches long, 27 inches high and 24 inches deep. It was recessed on the bottom to accommodate the entrance (Figure 31 and 32). Close inspection of the stone indicates it was installed in place and there are no indications it was added after the magazine was completed.

It is unknown who carved the sculpture as there are no mason identifying marks. The men who might possibly have been the artists are listed below. They were all involved with building the magazine during July to October 1857 when the stone was excavated from the quarry and the sculpture was created:

- Caleb S. Merrill, Master Mason
- Peter Larseneur, Master Stone Cutter
- David Gorman, Mason
- John Brislan, Mason
- John Buckley, Mason
- John Kelley, Mason
- John O'Conner, Mason

The likely sculptor of the eagle and cannon was Peter Larseneur. Larseneur was a marble and freestone cutter by profession. He possessed an artistic touch in designing decorative monumental work for cemeteries. His handiwork may be seen in the sculptured ornaments in stone that adorn the Catholic Cathedral on California Street in San Francisco.

John Gomo has been generally credited with having carved the sculpture and is listed in several resources as "reportedly," "supposedly," "reputedly," and "local tradition credits" John Gomo being the sculptor. Unfortunately, the historical timing of Gomo moving to Benicia makes this questionable.

John Elinor Gomo was born in France March 26, 1825. He immigrated to the

US in 1850 and married Catherine Smith, a native of Monaghan, Ireland, on October 9, 1852 in Albany, New York. John and Catherine had three children in New York, John Thomas (born 1853), Elizabeth (born July 1857) and Louisa (born February 1859).[54] Gomo came to Benicia with his family in May 1859, three months after Louisa was born.[55] John and Catherine had four more children in California: Maggie, Charles, Louis, and Frank.

The Benicia Arsenal ledger does not list Gomo as a hired worker until June 1859 where he was a quarryman working on the construction of the Clock Tower, not on the magazines (Figure 33).

So, how did this local tradition about John Gomo and the Powder Magazine start? In the April 7, 1949 issue of the *Benicia Herald-New Era*, there was a picture of Gomo's daughter, Mrs. J. E. (Maggie) Hobbie, pointing to the "Eagle and Cannon" on Building 10 claiming this was the handiwork of her father. Maggie Gomo was born in Benicia June 1860, two years after the magazine was built (Figure 34).

Gomo did work on the Clock Tower and continued to be employed by the Arsenal as a "mechanic," compensated at $5.00 per day, until at least 1875.[56] He lived in Benicia at a house on East I Street and East 7th Street at the top of the hill. He completed monuments in the Benicia City Cemetery; an example is the sandstone work on the obelisks surrounding the Goodyear grave site.[57] He became a naturalized citizen on February 6, 1861. John died in 1887 at the age of 61 and Catherine died in 1908 at the age of 79. They are both buried at the Benicia City Cemetery.

The eagle and cannon was not the only such sculpture locally created in 1857. The Mare Island Naval Shipyard, founded in 1854, was also building a powder magazine. The magazine, now called Building A-1, was made in 1857 from sandstone blocks from a quarry on Angel Island[58] and had a roof made from wood framing and steel, rather than a vaulted stone ceiling (Figure 35).

On October 17, 1857, Mare Island Naval Commander David Farragut noticed that his masons "placed a frontis stone on the magazine."[59] Over the entrance was a carved stone eagle clutching an anchor surrounded by a wreath that was hand hewn by Peter Kennedy (Figures 36 and 37).

Peter Kennedy was born in Ireland in 1828 and came to the USA in 1845 at the age of 18.[60] During the Gold Rush he panned gold in Tuolumne and Yuba counties.[61] He later moved to Vallejo, where he was a stone mason and helped construct the Mare Island Naval Shipyard magazine, the first Naval magazine in the West. In Vallejo, he lived with seven other stone masons from Ireland, New York and Washington DC.[62] He lived in San Diego from 1870 to 1875 and no records could be found after that date.

Figure 29: Eagle over Powder Magazine 10 entrance.
Source: Photograph by author.

BENICIA ARSENAL POWDER MAGAZINE

*Figure 30: Drawing of eagle over magazine entrance, 1976.
Source: Library of Congress, HABS Cal, 48-Beni Sheet 1.*

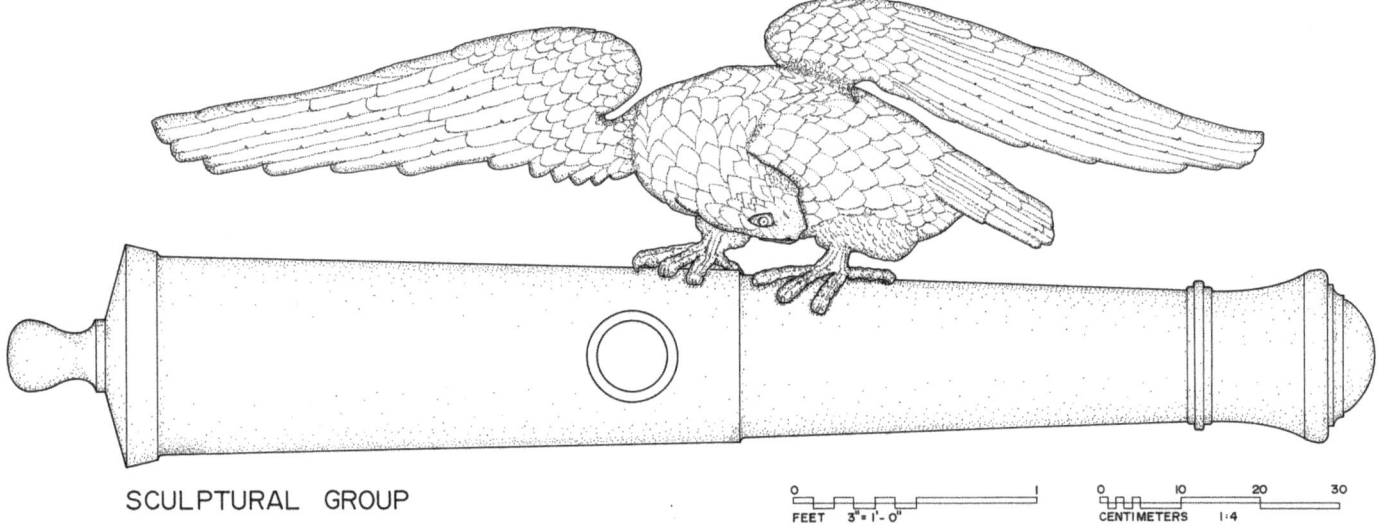

SCULPTURAL GROUP

THIS POWDER MAGAZINE, ARSENAL BUILDING NUMBER 10, WAS CONSTRUCTED IN 1857. IT IS ALMOST AN EXACT COPY OF THE FIRST STONE MAGAZINE AT THE ARSENAL WHICH HAD BEEN CONSTRUCTED THE YEAR BEFORE, AND WHICH ALSO SURVIVES. BOTH ARE MASSIVE GROIN-VAULTED STRUCTURES OF LOCAL SANDSTONE ASHLAR. BUILDING NUMBER 10, IN ADDITION, HAS ABOVE THE DOOR A FINE SCULPTURAL GROUP, SUPPOSEDLY EXECUTED BY FRENCH MASON JOHN GOMO, REPRESENTING AN EAGLE PERCHED ON A CANNON BARREL.

THIS PROJECT WAS UNDERTAKEN BY THE HISTORIC AMERICAN BUILDINGS SURVEY IN COOPERATION WITH EXXON COMPANY, USA, AND THE BENICIA HISTORICAL SOCIETY. UNDER THE DIRECTION OF JOHN POPPELIERS, CHIEF OF HABS, THE PROJECT WAS COMPLETED DURING THE SUMMER OF 1976 AT THE HISTORIC AMERICAN BUILDINGS SURVEY FIELD OFFICE, BENICIA, CALIFORNIA, BY JOHN P. WHITE (ASSISTANT PROFESSOR, TEXAS TECH UNIVERSITY), PROJECT SUPERVISOR; ROBERT BRUEGMANN (UNIVERSITY OF PENNSYLVANIA), PROJECT HISTORIAN; KENNETH PAYSON (CORNELL UNIVERSITY), ARCHITECT; AND STUDENT ASSISTANT ARCHITECTS SCOTT BARNARD (UNIVERSITY OF PENNSYLVANIA), JAMES L. COOK (TEXAS TECH UNIVERSITY), GARY A. STATKUS (UNIVERSITY OF ILLINOIS, URBANA).

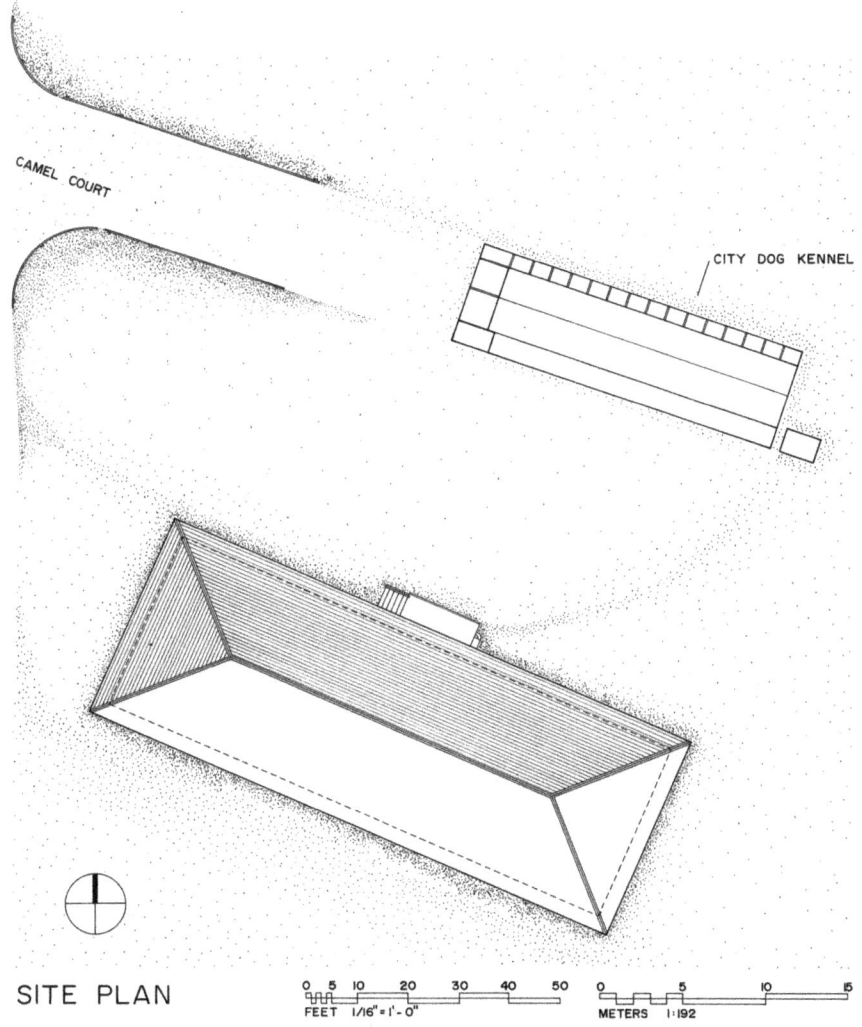

SITE PLAN

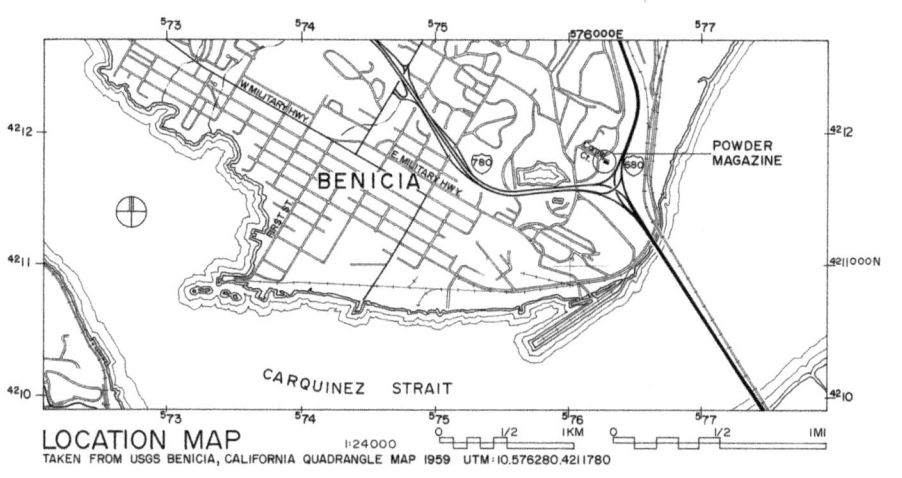

LOCATION MAP
TAKEN FROM USGS BENICIA, CALIFORNIA QUADRANGLE MAP 1959 UTM: 10.576280,4211780

Figure 31: Image depicting how eagle and cannon were sculpted from a single stone (outlined).
Source: Photograph by author.

Figure 32: Close up of Eagle.
Source: Photo by author.

Figure 33: John Gomo listed in the Arsenal ledger for the first time in June, 1859.
Source: Photograph by author.

Figure 34: Mrs. J. E Hobbie at Magazine 10.
Source: Benicia Herald-New Era, April 7, 1949.

Figure 35: Stone Magazine A-1 at Mare Island, 1998.
Source: National Park Service, HABS CAL, 48-MARI,1B-6.

Figure 36: Eagle and wreath sculpture on Magazine A-1, 1954.
Source: Library of Congress, HABS CA-1543-B-4.

Figure 37: The mark of P. Kennedy on lower right on carving, 1998.
Source: Library of Congress, Survey number: HABS CA-1543-B-4.

Chapter Ten

The Building of Stone Powder Magazine 10

In 1857, Captain Francis Callender, Ordnance Depot, Benicia Arsenal, hired Caleb S. Merrill, master mason, to build the new powder magazine. Merrill, having built the first stone magazine in 1855, was experienced and available. He also hired Peter Larsenuer, a master stone cutter who had worked with Merrill in the recent past.

The Benicia Arsenal ledger of monthly activities during this time provided a detailed account of the personnel, their trade, duties, quarry work and building construction.[63] In January and February 1857, rainfall of over 11 inches fell limiting the initial work.[64] A hired work crew of 5 men began excavating and removing loose rocks at the quarry located 700 feet east of the magazine site (quarry 2 in Figure 1). Quarry work was difficult and the laborers used the available tools of the trade: chisels, hammers, sledges, picks, and saws (Figures 38 and 39).

The construction of Magazine 10 began in March 1857 and it was well-planned and coordinated. The site was prepared by removing 9,519 cubic feet of dirt and rock. Rough sandstone blocks to be used for the foundation were removed from the quarry and cutting of the sandstone blocks began. In April, "dimensioned stone" (cut to specific size and shape) continued to be quarried and transferred to the stone cutting yard. Stones were cut for arches, pilasters (columns), door sills, and water table. A water table is a masonry feature, sometimes decorative, that consists of a projecting layer (course) that deflects water running down the face of a building away from the foundation.

By May 1857, excavating and grading of the magazine site was completed and the laying of the magazine sandstone foundation commenced. The foundation was a rectangle measuring 104 feet by 35 ½ feet made from sandstone and was 4 ½ feet wide by 4 ½ feet deep, most likely resting on sandstone bedrock. It needed to be perfectly level. This base made for a solid foundation on which the sandstone walls could be placed and allowed a space for adequate ventilation under the floor. At the same time, the six column foundations were also set into place, using a pyramid shaped base (Figure 40).

The grounds around the magazine and quarry were a busy place (Figure 41). Stones for pilasters, center piers, groin arches, quoins (decorative end stones), headers, and door frames were cut. The work crew in May now numbered 39, with activities continuing in the quarry, stone cutting yard and magazine site.

In June, the foundation of the new magazine was nearly completed, The workforce was a total of 63 men consisting of 8 masons, 7 stone cutters, 5 stone dressers, and 43 laborers. The primary activities were laying the foundation stone, quarrying stone for the water table, walls, pilasters, and arches, and cutting stones to specific sizes and dressing the "nobling" stones (the exterior, decorative side of the wall stones). Laborers worked in the quarry to haul away rock debris to other areas of the Arsenal to be used for grading. The blacksmith was constantly busy sharpening and repairing the workers' tools.

By July, the foundation was completed with 402 cubic yards of foundation stone laid in place. Then, building of the water table course and walls began. The walls consisted of a water table course, nine sandstone block wall courses and a ceiling course built upon the foundation. The walls alternated two and three stones thick for each course. This can be seen at the entrance (Figure 42). After the floor course, the sandstone blocks layers decreased in thickness as each course was laid (Figure 43). This was a common practice in stone masonry to increase stability.

The style of the masonry used in the magazine construction is known as "Flemish bond." This style has the outer facing stones laid lengthwise (stretchers) and end facing seemingly square blocks (headers) laid across the wall (Figure 44 and 45).

In August, the Army hired Merrill's son, Caleb Merrill Jr., as a stone dresser and quarryman. By the end of that month, 7000 cubic feet of stone walls had been laid and the windows and front entrance were completed. The stone stretcher over the entrance was the longest stone in the magazine measuring 10 feet two inches and the eagle and cannon must have been sculpted at this time. The workforce of 47 men, consisting of 8 stone cutters, 8 masons, 6 stone dressers, 6 first class laborers, and 19 second class laborers continued to work on the magazine and in the quarry.

The following month, 3000 feet of lumber was prepared for the roof. Another 2000 cubic feet of stone wall was laid and the installation of the ceiling arches commenced. A total of 170 ceiling arch stones (consisting of 1500 cubic feet) were readied by the masons and stone dressers.

The vaulted ceiling structure begins at the fifth course. A "vault" is created wherever arch shapes intersect.[65] This design is called a "groin vault," also known as a cross vault because the intersected sections resemble the shape of a cross. The main advantage of the groin vault is that it takes the weight of the roof and concentrates it on just four points at the corners of each column. The ceiling is composed of wedge-shaped pieces called "voussoirs" which are held in place, like the stones in an arch, by the pressure of the neighboring pieces.

Two possible methods could have been used to hold the ceiling stones in place during construction:

- A curved wooden form called a "falsework" that was placed underneath in the shape desired and then each stone is stacked against each other up the curve to meet at the top where a tapered keystone block was placed and the wood falsework removed.
- The interior of the building was filled with a material such as sand to support the ceiling blocks until the point where the keystone blocks could be put into place.

Each stone voussoir was perfectly hewn to fit in place. In the ceiling at the center of the arches are the keystone blocks, totaling 14 in number. They are in a cross pattern, each measuring 28 inches across the cross and 12 inches on each arm. The keystones and the stones in line with the crosses are perfectly horizontal across the ceiling structure (Figures 46). Handles were affixed to the keystone to allow for lifting and setting into place. The arch is stabilized by the weight of the stones to keep the shape with a tight, compact fit. There are two rows of seven vaulted areas in this structure. No bonding materials or cement were used. Figure 47 shows the completed interior vaulted ceiling and column structure and Figure 48 shows an isometric drawing of vaulted ceiling and columns.

By the end of October 1857, the magazine walls and part of the plinth course were finished. The plinth course is a projecting course of stones around the wall foundation and column foundations that provide for a more even distribution of weight from the following courses. The groin arch installation was nearly finished and 140 boxes of slate were purchased most likely from the Eureka Slate Company in Slatington, California. A slate mason was hired to prepare slate for the roof in which individual pieces were dressed 7 inches by 16 inches each with two nail holes. The timbers for the roof were framed and 100 feet of copper was used for the roof gutter. Copper bolts for the magazine door and copper nails for the roof were made by the blacksmith.

In November, the rains returned and work force of 32 men finished installing the final 600 cubic feet of groin arch ceiling stone. In December, the final course of plinth stone was cut and laid to complete the magazine's stone work. An additional 150 feet copper gutter and eaves were manufactured.

In January 1858, the work on the roof of the magazine was nearly finished and about one third of the slate had been laid. The following month, the slate roof was completed. At this time, construction began on a cistern at the west end of the magazine. Brick-sized stone was cut and dressed and 10,000 were laid in the cistern walls.

In March, the platform at the magazine entrance and flooring were completed. The floor consisted of 4 ⅛ inch tongue and groove planks running the length of the magazine on underlying beams, supported by the sandstone foundation. The interior floor measures 95 feet 10 inches by 25 feet 7 inches.

In the following months, the finishing touches were made to the interior. The wainscoting was installed and painted and the gaps between stones were sealed. The magazine was completed. The grounds around the magazine were leveled and graded. The last course of stone was cut and laid in the walls of the cistern.

A total of 90 men were involved in construction of Magazine 10 which took 14 months to complete. The nationality of the workers is listed in Table 2.

Table 2: Nationality of powder magazine workers

Nationality	Total Men	Percent
Irish	46	51
USA	11	12
British	5	5.5
Scottish	4	4.5
Canadian	3	4
German	2	2
Swedish	1	1
Welsh	1	1
French	1	1
Unknown	16	18

There was a total of 8,855 man-hours utilized in the construction and 3,024 stones of different sizes were utilized. The breakdown was:

- Foundation and water table courses: 230
- Walls: 2,060
- Columns and ceiling: 734

Building of the magazine was tough work for the hired workers. Many lived in the Benicia Barracks and some lived in the city of Benicia. David Gorman, age 32 and a native of Ireland and a stone mason, seemed to be an Army favorite as he was usually listed in the beginning of the ledger each month and was also involved with the building of the Clock Tower. He enlisted in the Army and lived in Benicia with his wife Leonora and their seven children. David died at the age of 58 in 1884 and is buried in the Benicia City Cemetery.

Robert McBride, a native of Ireland, joined the Army under Captain Stone on November 15, 1853. He performed duties as a blacksmith, carriage maker, hospital assistant and guard. He was discharged October 11, 1856 due to expiration of enlistment but was retained as a hired laborer during construction of the magazine. McBride re-enlisted with the Army on December 1, 1857 for five years. Robert and his wife Mary lived in Benicia and when their son John was of age, he worked in the Benicia Tannery. All three lived their entire lives in Benicia and are buried in the City Cemetery.

Eleven other men, Michael Cain, Thomas Dowling, Patrick Fitz Patrick, John C. Gordon, Thomas Miller, Thomas Moran, William McDonald, John McMullin, John Pattison, James Smith, and Peter Larney Smith enlisted in the Army and continued working at the Benicia Arsenal.

Sometimes tragedy caught up with the workers. James Kelly, a mason, is listed in the Army ledger by Ordnance Captain Franklin Callender as "died June 16, 1857." There is no mention of how he passed, he had only worked 7 days in April and 1 ½ days in June. Another tragedy befell Patrick Gardiner who was a hired blacksmith. On August 25, 1857, he was attempting to board the *New World* stream ship in Sacramento to travel to San Francisco when he fell down a stairway, broke his neck and died. He apparently was intoxicated and did not have a ticket. The clerk was accused of giving him a shove causing him to fall.[66] He was age 46 and buried in the Sacramento City Cemetery.

70 BRUNNER & LAY

QUARRY OR BREAKING HAMMERS

10 to 24 lbs., hand made, per lb. 35c

QUARRY SLEDGE

14 to 20 lbs., hand made, per lb. 26c

SPAULDING HAMMERS

Spaulding Hammers, 10 to 12 lbs., per lb. 25c
" " 12 to 14 lbs., per lb. 20c
" " 14 to 16 lbs., per lb. 25c

All Hammers are our own hand make

MASONS' HAMMERS

Masons' Hammers, 6 to 7 lbs. $1.75
" " 7 to 8 lbs. 2.00
" " 8 to 9 lbs. 2.25

570 WEST POLK STREET 71

STONE PICKS

Stone Picks, 6 to 14 lbs., per lb. 18c
Made of Solid Cast Steel.

QUARRY PICKS

Quarry Picks, 6 to 14 lbs., per lb. 18c
Made of Solid Cast Steel.

BULL POINT FOR STONE OR GRANITE

Bull Points, 6 to 10 pounds, per pound 20c

STEEL FORGED CAISSON AND TUNNEL DIGGERS' GRUBS

Weight, 7 lbs.; 11 to 12 in. long over all. Face, 3-inch, each $2.00

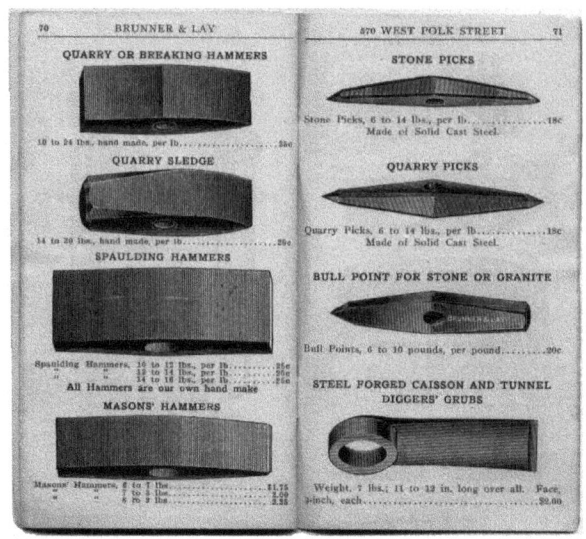

Figure 38: Quarry tools in Brunner & Lay catalog.
Source: Brunner & Lay Tools, Chicago, Illinois (date unknown).

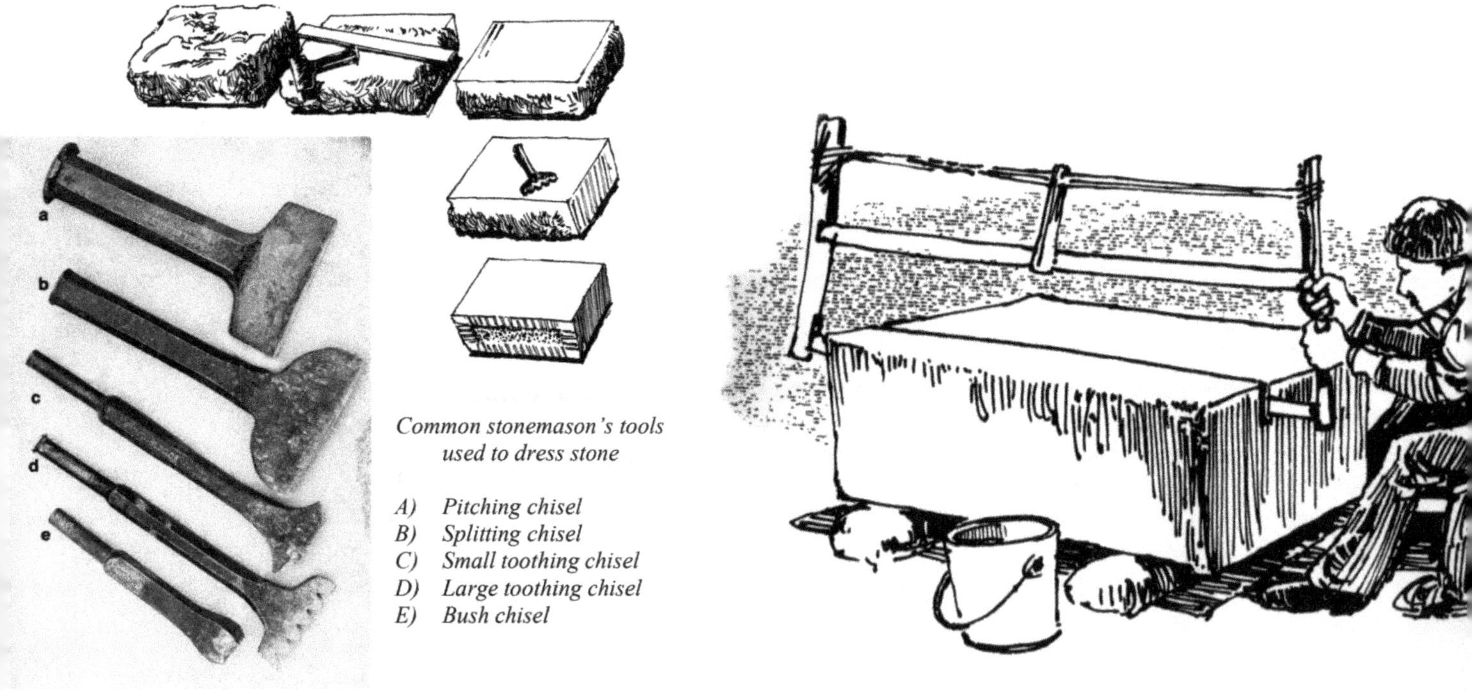

a
b
c
d
e
Common stonemason's tools
used to dress stone
A) Pitching chisel
B) Splitting chisel
C) Small toothing chisel
D) Large toothing chisel
E) Bush chisel

Figure 39: Quarry tools and sawing stone.
Source: Kenton County Historical Society, Kentucky, July-Aug 2015.

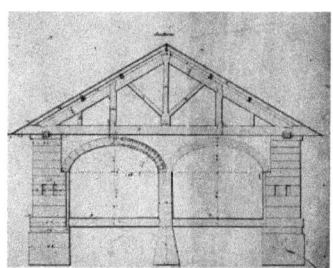

Figure 40: Powder Magazine cross section drawing.
Source: Library of Congress, HABS CA-1948 image 10.

Figure 41: Typical scene of what the quarry activity may have looked like.
Source: Edwin Bell Howe, Deep Lock Quarry, Ohio.

Figure 42: Wall thickness construction at entrance.
Source: Photograph by author.

Figure 43: Diagram of course sizes.
Source: Photograph by author.

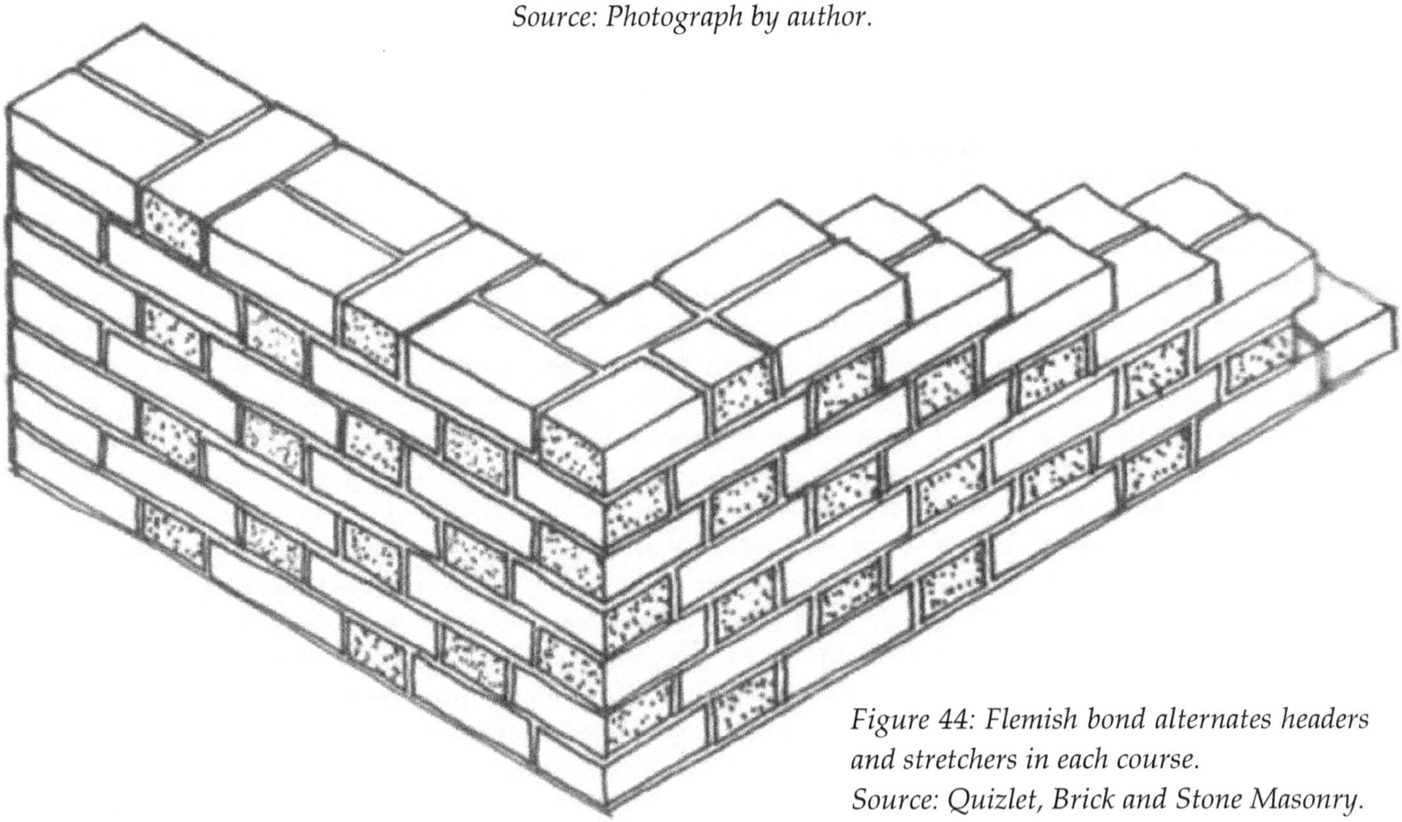

Figure 44: Flemish bond alternates headers and stretchers in each course.
Source: Quizlet, Brick and Stone Masonry.

Figure 45: Magazine wall Flemish style of headers and stretchers.
Source: Photo by author.

Figure 46: Keystone block and supporting stones.
Source: Photograph by author.

Figure 47: Josephine Cowell in Magazine 10, 1955.
Source: Library of Congress, HABS no. CA-1948-3.

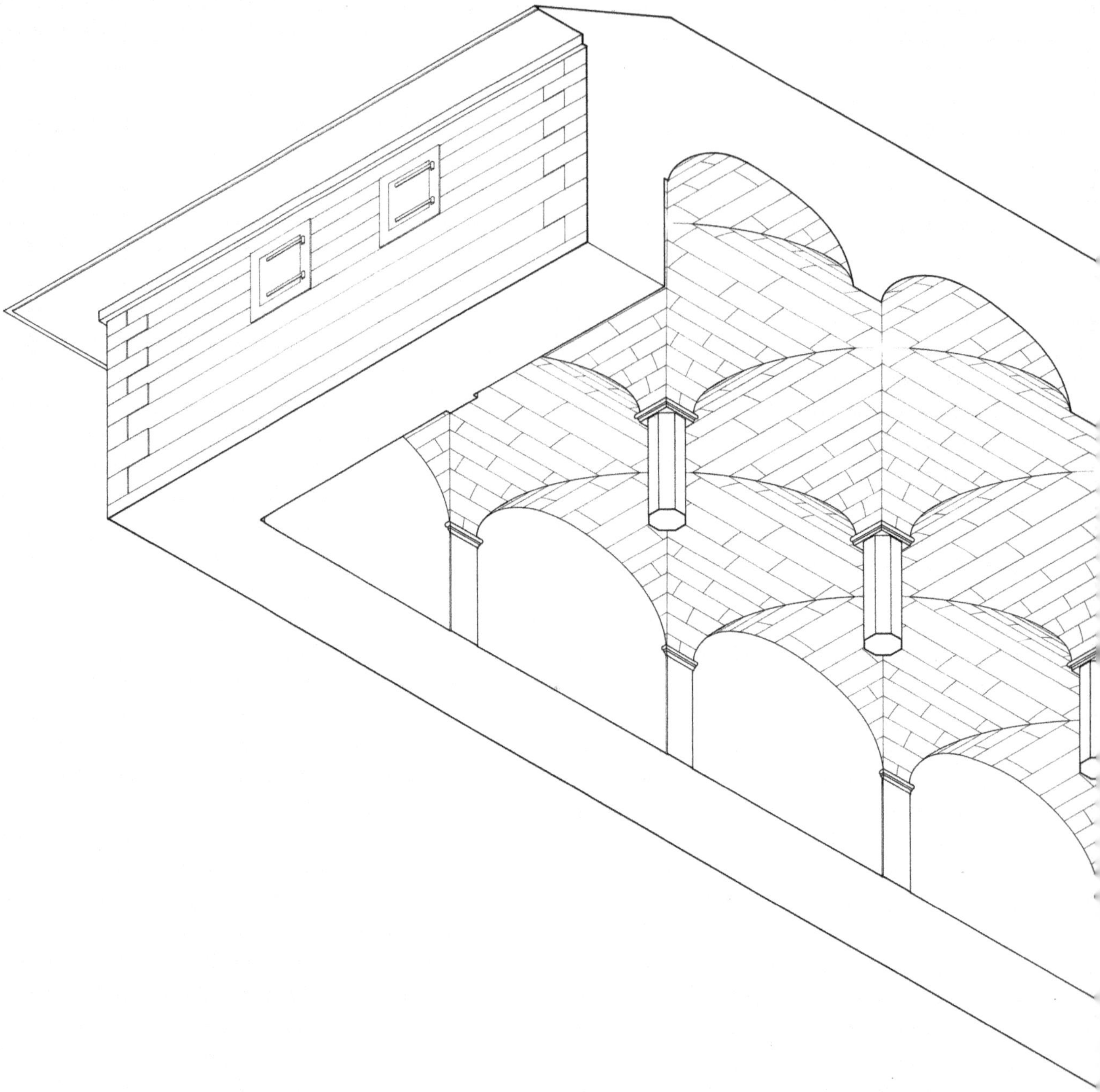

Figure 48: Isometric drawing of vaulted ceiling and columns, 1976.
Source: Library of Congress, HABS Cal, 48-Beni Sheet 5.

ISOMETRIC VIEW OF INTERIOR VAULTING

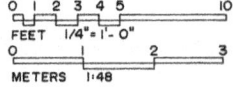

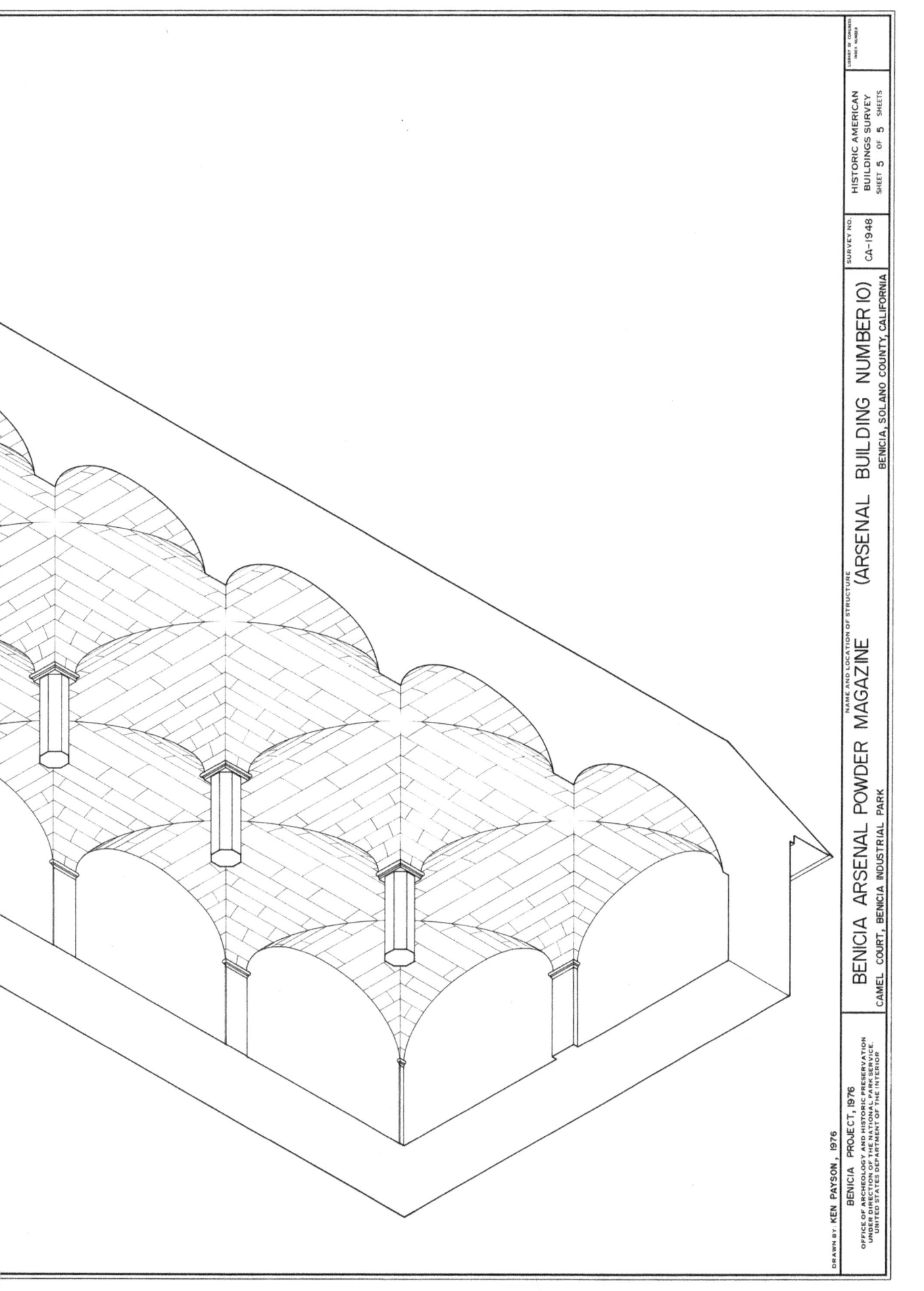

Chapter Eleven

The Stone Powder Magazine 10 Exterior

The powder magazine has two windows each on the building west and east sides. They measure 32 inches by 32 inches and were located between wall layers 5 through 8 (Figures 49 and 50).

There are two vent holes (12" x 6") on the north and south sides of the magazine on wall level 5 which are covered by copper plates with a series of small holes (Figure 51). Around the foundation of the powder magazine there are small holes measuring 9 inches by 9 inches which provide ventilation beneath the wooden floor, in a maze of sandstone block supports. There are four each on the building front and back and three each on the sides (Figure 52).

There are several inscriptions engraved by Arsenal personnel over the years on the floor level sandstone on the east side (Figures 53 and 54).

Adding to the beauty of the exterior design, the entrance and corners have stylized "quoins" (Figures 55 and 56). Quoins are masonry blocks at the corner of a wall. These imply strength, permanence, and expense, all reinforcing the onlooker's sense of a structure's presence.[67]

The entrance to the powder magazine has a large outside door (Figure 57). It measures 58 inches by 94 inches and is 2 ½ inches thick. It is made from panels of "copper-clad steel" manufactured by the US Steel Company (formerly Carnegie-Illinois Company) in the late 1930s (Figure 58). The reason for the copper-clad is that it covers the steel that could generate a spark if struck by an object and copper will not produce a spark. The magazine walls are four feet three inches thick. This can be seen at the entrance in Figure 59.

Figure 49: The powder magazine windows and foundation vents.
Source: Photograph by author.

Figure 50: Exterior window with door.
Source: Photograph by author.

Figure 51: Exterior vent hole and closeup of copper plate with holes. Source: Photograph by author.

Figure 52: Foundation vent holes.
Source: Photograph by author.

Figure 53: Inscription from "2nd Cal Inf. August 1917" and "JM 1924".
Source: Photograph by author.

Figure 54: Inscription from "Bales 1932" and, "Tex Kid Windy" April 1942, Source: Photograph by author.

Figure 55: Stylized entrance and corner quoins.
Source: Photograph by author.

Figure 56: Closeup of corner quoins.
Source: Photograph by author.

Figure 57: Outside door (the box in the center is a modern light).
Source: Photograph by author.

Figure 58: US Steel Company inscription on door panel:

FIRE DOOR STANDARD
1C-20-LB.
COPPER STEEL
CARNEGIE - ILLINOIS
Source: Photograph by author.

Figure 59: Sandstone walls at entrance (4' 3" thick).
Source: Photograph by author.

Chapter Twelve
The Powder Magazine 10 Interior

The magazine interior doors provide protection, and humidity and temperature modulation. They are made from wood and are 2 inches thick (Figure 60). The interior consists of one large room divided by a row of columns (Figure 61). The exquisite, patterned, vaulted ceiling and modified Corinthian pillars were hand-fashioned and placed by the masons.[68] Splendid stylized seashells (reminiscent of the Greek and Roman palmettos) decorate the Doric capitals of the pillars.[69] The vaulted ceiling was practical as well as aesthetic; it was designed to divert the effects of any explosion, upward, rather than sideways.

Why the builders chose to create a beautiful interior with such splendid carvings, only to be used as a place to store gunpowder, is unknown.[70] The columns are decorated with carved stylized shapes, seemingly a pendant, a seashell and a scroll. The carvings are in pairs: each of the two east, center and west columns have the same patterns (Figure 62). The wider decorative protruding stone above the carvings is called a "capital." In the four corners, only a small decorative portion of the capital is visible (Figure 63).

The interior walls have wooden wainscoting. An interior framework was constructed using 3 inch by 4 inch redwood beams against the sandstone block walls in which wooden planks, 3 ½ inches and 5 ½ inches in width by 7/8 inch thick, were attached (Figure 64).

The four interior windows each have inside doors. For lighting, smokeless oil lamps were placed in these window air passages (Figure 65). There are two small vents in the interior on the north and south walls measuring 12 inches by 6 inches (Figure 66).

Red lines on the floor indicate location of shelving racks that once stood there (Figure 67). The shelving was limited to a load of 100 pounds per square foot.

For nearly one hundred years, the Benicia Arsenal was used primarily for ammunition, powder, and ordnance storage and for the manufacturing of black powder. The amount of black powder manufactured yearly ranged from 150,000 to 250,000 pounds, and the total amount stored in 1905 was approximately 500,000 pounds.[71]

Black gunpowder was stored in small white oak barrels measuring 18 inches diameter and 20.5 inches long and each held 100 pounds of powder. This magazine was built to store up to 3,000 barrels. The magazine floor layout consisted of racks on both sides of the columns and along the west wall. The racks were 56 inches wide and the aisle ways were 32 inches wide. Based on the capacity, the shelves must have been about 4 feet high (Figure 68).

Near the door entrance and on some of the columns, there is a black tar-like substance. This substance was most likely used for sealing the powder kegs (Figure 69). Wall signs showing rack location for inventory and floor load limit (Figures 70 and 71).

There are many wall inscriptions of names, initials and dates of Arsenal personnel. Many other markings of interest include name of the Arsenal, numbers and other interesting graffiti (Figures 72 to 80). The oldest two dated inscriptions are from 1874 and 1878. A near complete list is given in Appendix B.

Walter Kührt was born in Zella-Mehlis, Germany, November 25, 1908. He entered the German army in 1942 and was with the 1st Company Replacement Rifle Battalion 471, stationed in Butzbach, Germany. In 1943, he was with the Grenadier Regiment 765 unit, and later he served as a Corporal in the 242nd Division. Walter was captured on August 15, 1944 in France and was assigned prisoner number 81G 345841H. He was detained in the following camps in America[72]:

- Camp Florence, Arizona, September 1944
- Camp Haan, California, April 1945
- Camp Stockton, California, May 1945
- Camp Cooke, California, November 1945
- Camp Stockton, California, February, 1946
- Benicia Arsenal, California, March, 1946

While in the Benicia Arsenal, he was working with the US Army and on March 28, 1946, he wrote his name on the wall of Powder Magazine 10.

Figure 60: Inside doors.
Source: Photograph by author.

Figure 61: Interior columns of Powder Magazine 10, 1948.
Source: Benicia Historical Museum.

Figure 62: Detail of east, center and west interior columns.
Source: Library of Congress, HABS no. CA-1948 Images 5, 6 and 7.

Figure 63: Capital in corner.
Source: Photograph by author.

Figure 64: Redwood 3 inch by 4 inch beams were the infrastructure of the interior wainscoting. Note the rough finish of the interior face of the sandstone block. Source: Photograph by author.

*Figure 65: Interior window.
Source: Photograph by author.*

Figure 66: Interior vent.
Source: Photograph by author.

Figure 67: Floor lines marking location of racks.
Source: Photograph by author.

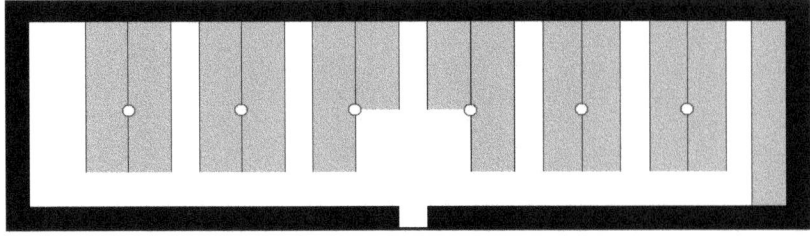

Figure 68: Powder magazine floor layout, entrance at bottom, shaded areas are rack locations. Source: Diagram by author.

Figure 69: Black tar sealant on wall column. Source: Photograph by author.

Figure 70: Inventory location signs on south wall, the light in the center is modern. Source: Photograph by author.

Figure 71: Load limit sign on west wall. Source: Photograph by author.

Figure 72: Benicia Arsenal inscription.
Source: Photograph by author.

Figure 73: Math wall markings, possibly adding weight measurements.
Source: Photograph by author.

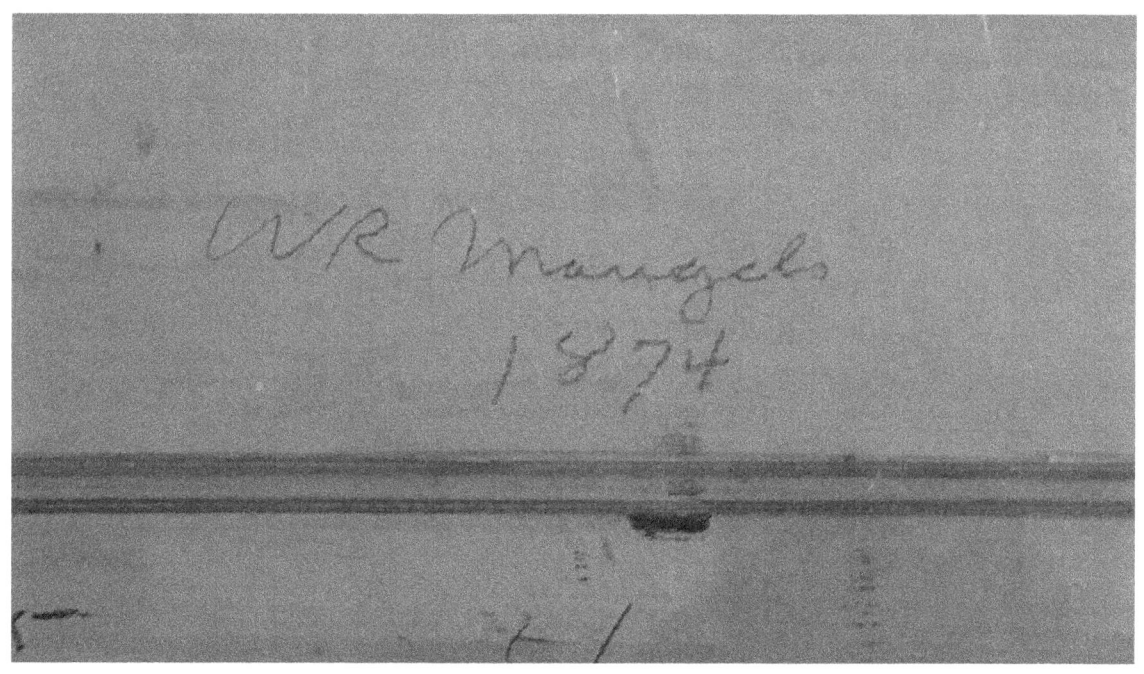

Figure 74: Oldest wall inscriptions
"W. R. Mangles 1874" and "William J. Flannery July 9th 1878."
Source: Photograph by author.

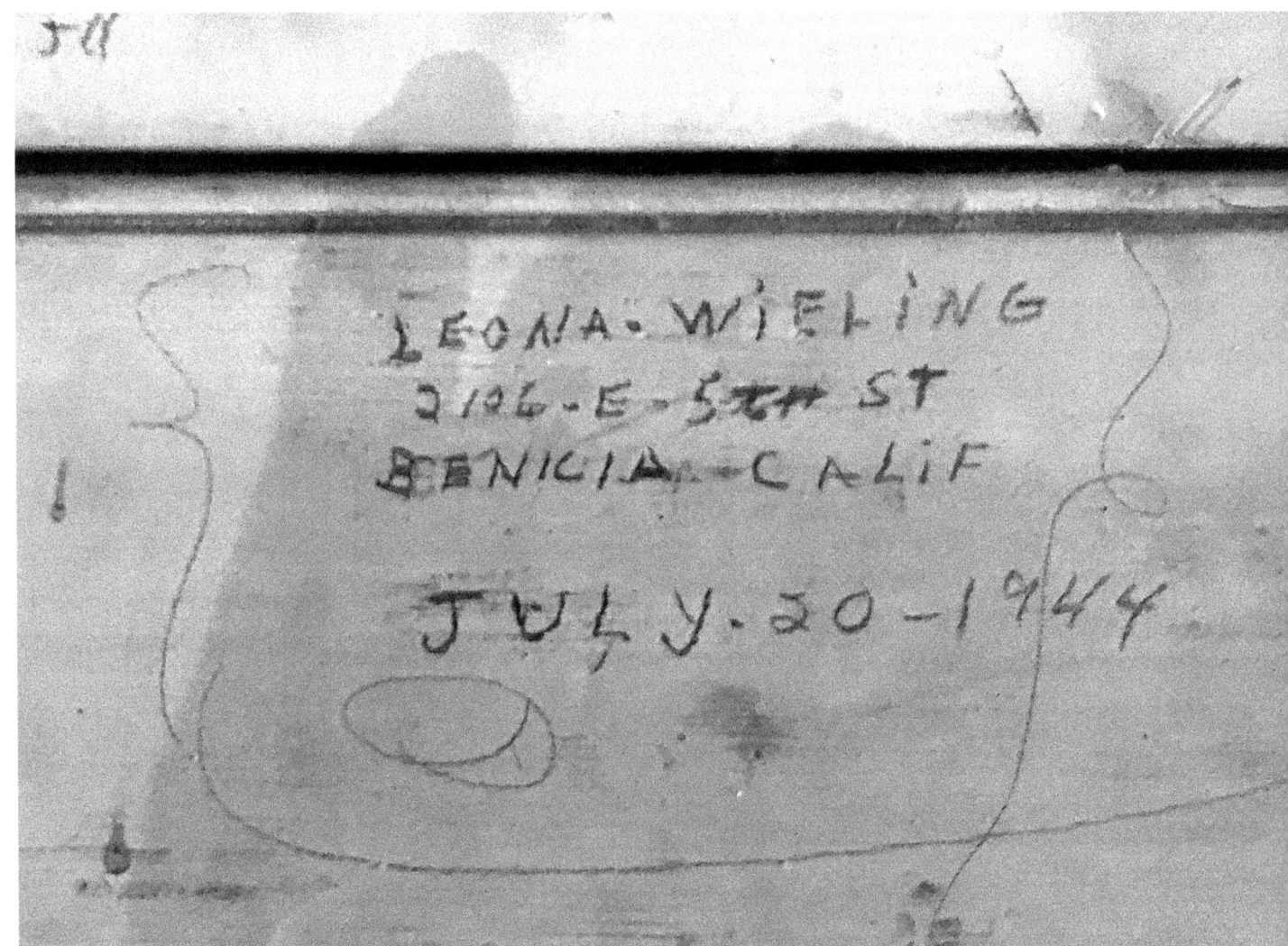

Figure 75: One of the few woman to list her name:
"Leona Wieling 2126 E. 5th Street Benicia, Calif July 20, 1944."
Source: Photograph by author.

Figure 76: "Alice O' C: on wall under column.
Source: Photograph by author.

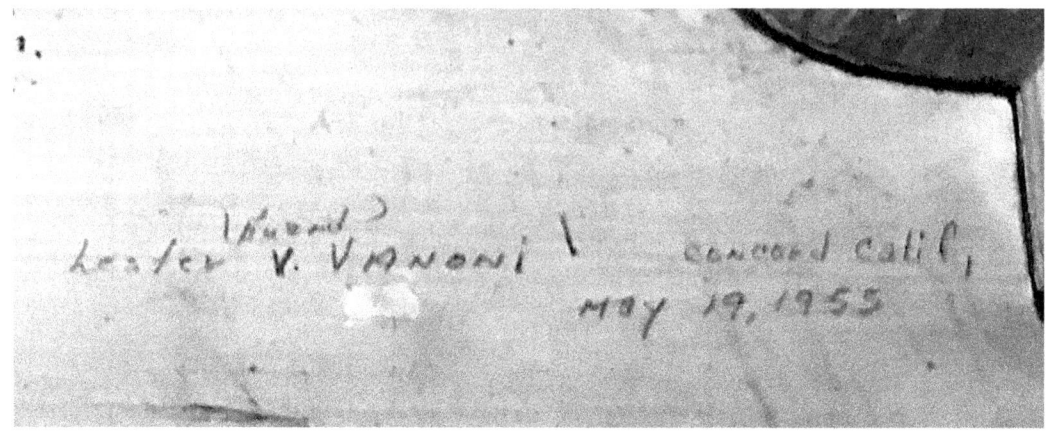

Figure 77: "Guard Lester V. Vanoni, Concord, CA May 19, 1955."
Source: Photograph by author.

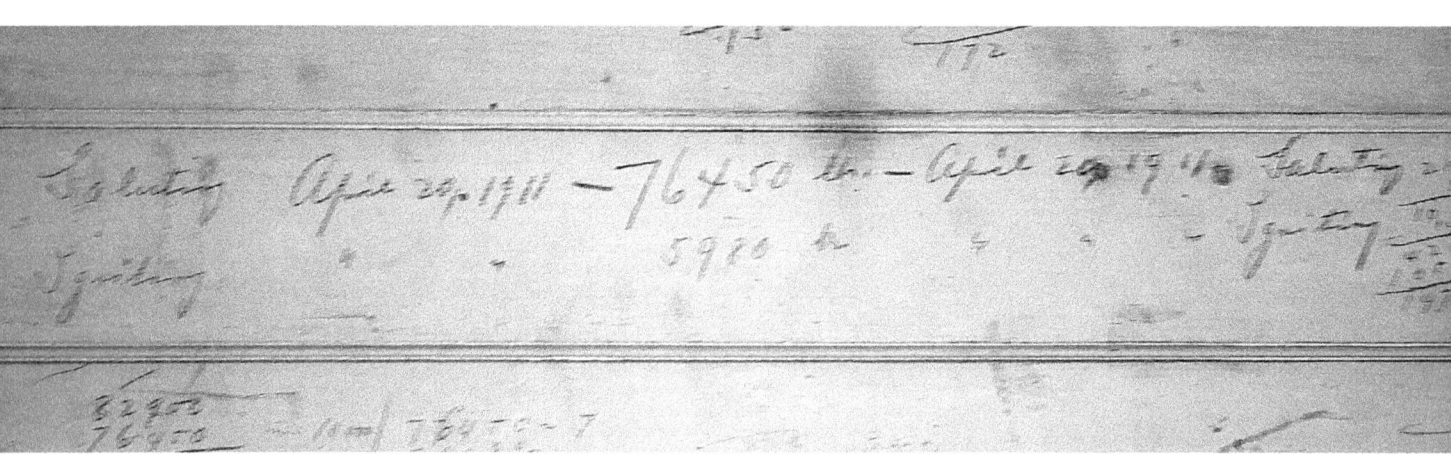

Figure 78: An interesting inscription: "Saluting" "Igniting" April 20, 1911.
Source: Photograph by author.

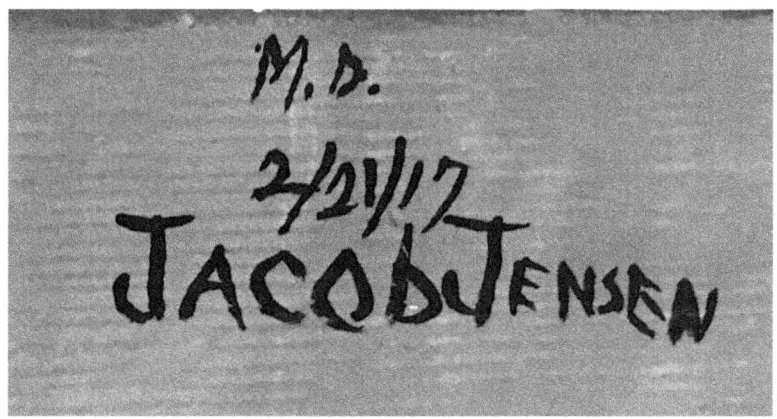

*Figure 79: "Jacob Jensen, M.D., 2/21/17"
used black tar on southwest column corner.
Source: Photograph by author.*

*Figure 80: "Walter Kührt 28-III-1946" (German prisoner-of-war).
Source: Photograph by author.*

Chapter Thirteen

The Fire of 1922

On August 13, 1922, 15 employees of the Columbia Salvage Company of New York contracted by the Army were working at a brick magazine located approximately 700 feet north of the Camel Barns. This magazine was used as an ammunition salvage warehouse for cartridges, grenades, shells, and bombs. In the afternoon they were cleaning powder from shrapnel shells when one worker attempted to empty a shrapnel shell by pounding it against a steel rail. A spark generated and quickly ignited loose powder in the surrounding area. The fire engulfed the magazine which exploded and was completely destroyed. The explosion hurled shrapnel shells over several hundred acres of ground, exploding as they descended.

Fire erupted all over the Arsenal grounds fueled by the dry grasses. The fire was brought under control about 7 pm through the efforts of combined fire departments of Benicia, Mare Island and Vallejo including 37 soldiers stationed at the Arsenal, 200 Marines from Mare Island, and civilian volunteers. *The San Francisco Examiner* captured the event the following day (Figure 81).

The text of the *San Francisco Examiner* August 23, 1923 newspaper article[73] is as follows:

POWDER MAGAZINE AT BENICIA BLOWS UP
$600,000 Loss in Blast at Arsenal Fire
Flames, Started by the Explosion of Shrapnel Shell
at Benicia, Destroy Ammunition
Laborer Who Caused Blaze Fatally Burned;
TNT Storehouse Has Narrow Escape

> With a detonation which shook Solano county towns, powder magazine No. 1 at the government army arsenal at Benicia exploded yesterday afternoon in a fire which swept over several hundred acres of the reservation.
>
> Puffs of smoke coming from the building a few minutes before the blast caused firefighters, including Capt. L. M. Wheeler, ranking officer at the post, to flee from the vicinity of the structure in time to escape fragments of the magazine which rained from the skies.

The fire was brought under control shortly after 7 o'clock last night through the efforts of combined fire departments of Benicia, Mare Island and Vallejo, 37 soldiers stationed at the arsenal, 200 marines from Mare Island and civilian volunteers.

The loss was estimated at $600,000.

TNT. BLAST AVERTED

Chemicals thrown around a magazine stored with TNT. checked flames rapidly before a heavy wind through stubble. Explosion of this magazine, officers said, would have wiped out the arsenal, caused a heavy loss of life and probably would have damaged the town of Benicia two miles away.

Several other magazines stored with high explosives were damaged and only saved with the with the greatest difficulty by the heroism of army and civilian fire fighters who worked under heavy a fire of shrapnel bullets from the exploding shells.

The fire is said by army officers to have been caused by the attempt of a Mexican worker to empty a shrapnel shell by pounding it against a steel rail.

A spark ignited loose powder and in a moment 1,000 three-inch shells, condemned by the army officers and sold to the Columbia Salvage Company of New York, were being hurled over several hundred acres of ground, exploding as they descended.

MAGAZINES MENACE

Fire immediately broke out in scores of places at once, menacing all twelve magazines on the reservation.

A fire apparatus from Benicia, two miles from the arsenal, arrived in time to be used in extinguishing the fire.

A second piece of apparatus from Vallejo arrived, but was not used.

The arsenal at Benicia is said to be the only government arsenal west of the Mississippi. Just recently it was heavily stocked with ordnance and high explosives.

The shrapnel being handled by the salvage company had been condemned by army officers and been sold. The employees of this company, all civilians, were emptying the shells of the powder when the accident occurred.

MAN RESPONSIBLE DYING.

Jose M. Fino, employee of the Salvage company, is the man whose action caused the conflagration, according to Army officers. He is in the emergency hospital at Vallejo in a dying condition. E. Narvaez, a cousin of Fino, assisted the man out of the danger zone.

Fourteen other employees of the Salvage company, engaged with Fino in handling the shrapnel, fled for their lives, the bullets kicking up the dust around them as they ran.

Some of these workers were still missing last night, but got off the reservation safely, Capt. Wheeler stated.

Mrs. Rudolph Quandt, the wife a sergeant stationed at the arsenal, was aroused at her home near the reservation by the noise of the explosion. She telephoned to the fire departments at Benicia and Vallejo for assistance and asked that ambulances be sent.

SHRAPNEL HITS BUCKET.

In the meantime she seized a bucket of water to extinguish the flames on the reservation near her home. As she climbed the fence a bullet from exploding shrapnel smashed the bucket and spilled the water.

While assistance was being rushed from Benicia and Vallejo fire departments, the thirty-seven soldiers stationed at the arsenal fought the flames at the points where they threatened to reach the most dangerous magazines.

To add to the menace of the high explosives, the shrapnel shells, flung to every point of the reservation, continued to explode at intervals, spraying bullets in all directions.

Missiles repeatedly struck the pumping machine manned by Private John Martin, but the soldiers stuck to their posts.

A large coal house was destroyed, together with the gun carriage store house and several smaller buildings, while the soldiers centered their attention on the magazines.

Late in the afternoon, it was believed that the flames had been brought under control.

Then puffs of smoke were noticed coming from Powder Magazine No. 1.

MAGAZINE EXPLODES.

When this exploded flames were again scattered to all parts of the reservation and the fight to save the other arsenal magazines started anew.

Two hundred marines and additional apparatus was hurried from Mare Island and guards were stationed to keep civilians away and to turn back traffic on a state highway skirting the reservation.

A.B. Sanders, secretary of the Benicia Chamber of Commerce, was fighting a fire which had started in the home of Sergeant C. M. Whittle when he noticed the flames attacking the TNT magazine.

The newspaper article refers to Magazine 1 blowing up. This was a brick magazine built in the 1850s. The powder magazine building 2 sustained roof damage by fire and flying shrapnel from this event (Figures 82 to 84).

Figure 81: The San Francisco Examiner August 13, 1922.
Source: San Francisco Examiner.

Figure 82: Site of the powder magazine explosion 1922.
Source: Benicia Historical Museum.

Figure 83: Explosion damage; Guard House at top center, Magazine 2 at left, 1922.
Source: Benicia Historical Museum.

Figure 84: Roof damage (arrow) to Magazine 2, 1922.
Power poles can be seen along front of magazine.
Source: Benicia Historical Museum

Chapter Fourteen
Saving Powder Magazine Building 10

In 1961, the US Defense Department announced the closure of the Benicia Arsenal on March 31, 1964, despite a renowned record of achievements for 112 years. The tasks of the Benicia Arsenal were to be divided between the Tooele Ordnance Depot in Utah and the Mt. Rainier Ordnance Depot in Washington State. At the close of the Arsenal, the property was sold to the City of Benicia for $4.6 million. The modern Arsenal buildings were turned into an industrial park and in 1974 the historical buildings that are part of the Benicia Historical Museum complex were kept by the city during a land exchange.[74]

The Magazine 10 sat idle for years after the Arsenal closure. The Benicia Arsenal was placed on the National Register of Historic Places November 7, 1976. The Benicia City Council dedicated the Camel Barn Museum complex on December 7, 1982. The Camel Barns and Powder Magazine were restored by the museum and was formally dedicated on May 19, 1985.

In the 1990s, the California Transit Authority (CalTrans) proposed to construct a new bridge across the Carquinez Strait either east or west of and parallel to the existing Benicia-Martinez Bridge and Southern Pacific Railroad bridge. A new toll plaza facility south of the strait and improvement of the bridge approaches north and south of the Carquinez Strait were also proposed.[75]

Selection of the west side bridge alternative would have required the demolishing of the Powder Magazine Building 10 and taking up to 5.9 acres of land from the Benicia Arsenal Historic District. Section 4(f) of the Department of Transportation Act prohibits use (taking) of any land from a historical site unless there is no feasible and prudent alternative to the use of such land.[76]

In 1992, Harold F. "Harry" Wassmann wrote a letter to CalTrans asking to save the Powder Magazine 10. Harry, a true Benician, was born in the family home that his grandfather had built in 1899 and spent his whole life there. Harry was a civil service employee who worked at Travis Air Force Base for most of his career. He retired after over 40 years of service and began his second career preserving the history of Benicia. A founding member of both the Benicia Historical Society and the Benicia Historical Museum, he served as president of both organizations. Harry was also responsible for promoting the Memorial Day service at the Benicia Arsenal Post Cemetery. He was proud of his town and its place in California history and he was generous in sharing his knowledge with others (Figure 85).[77]

The following is the content of his letter:[78]

> Benicia, CA
> February 28, 1992
> Mr. Preston W. Kelley CalTrans P.O. Box 7310, San Francisco, CA
>
> Dear M. Kelley:
> When you are considering the routes for the development of the Benicia- 1 Martinez Bridge Project, please do not select the alternative which would jeopardize the historic and architecturally unique Powder Magazine.
> The Powder Magazine is located in the Benicia Industrial Park and is one of the four buildings which comprise the Benicia Camel Barn Museum complex. The Powder Magazine is in excellent structural condition. It currently contains museum exhibit which are open to the public, and relate to the Benicia Arsenal.
> The Powder Magazine was constructed in 1857, one of the first permanent structures at the Benicia Arsenal. The Benicia Arsenal was a federal government installation (U.S. Army, Ordnance) established in 1851 and was the principal ordnance distribution depot west of the Rocky Mountains. The Powder Magazine was designed to hold 3,000 barrels of gun powder for national defense. It is constructed of native Benicia sandstone, quarried within a few yards of its location, finely processed and erected by stone cutters and masons of "old world" experience. It was constructed so that it would be almost impregnable. The walls are four feet thick. It has a heavy hipped roof and a vaulted ceiling of sandstone. On the front of the structure are small holes connected with horizontal passages in the walls which were apparently used for ventilation. The interior consists of a large room divided down the center by a row of columns supporting a series of cross vaults. Decorative carving in the sandstone of each column capital has been expertly executed as is, also, the sandstone sculpture over the entrance to the building - - an eagle perched on the barrel of a cannon. There is no other structure like it in the country.
> In its early years, the Powder Magazine was the source of gun powder for government troops manning fortifications along the Pacific Coast and numerous posts and camps in Washington, Oregon, California, Nevada and Utah. During the Civil War ordnance

materiel was shipped from the Benicia Arsenal to Atlantic Coast arsenals in support of the Union. The Benicia Arsenal equipped Civil War volunteers from Oregon, Nevada and California. Throughout the years - - years of peace - - and wars - - Spanish-American, World Wars I and II and the Korean War, the Arsenal performed its various missions with efficiency, pride and honor until its closing in l964.

I urge that you protect this irreplaceable structure for the benefit of students of history and architecture: and for those who cherish their local and national heritage.

Very truly yours,
Harold F. Wassmann

In late 2001, construction began on a newer bridge east of and parallel to the railroad bridge and Powder Magazine Building 10 was saved.

Photo 85: Harry Wassmann (1919 to 2017).
Source: Vallejo Times Herald.

Chapter Fifteen

The Powder Magazine Vaulted Ceiling History

The vaulted ceiling design of powder magazines dates back to colonial days. A good example is the Old Powder Magazine in Charleston, South Carolina. In 1703, the English Commons House of Assembly authorized construction of a brick powder house inside Charleston's fortification wall. The sum of £50 ($65) was allotted from the public treasury to build a powder magazine which was completed in 1713. It is a small brick building with walls three feet thick and the interior features eight 27-foot vaulted pillars around the perimeter and one in the center that were designed to implode in case of an explosion. The Old Powder Magazine on Cumberland Street is the oldest public building in Charleston (Figure 86).[79]

Other powder magazines in the U.S. which used a vaulted style ceiling are listed:

- Charleston Powder Magazine, South Carolina, built 1712
- Fort Matanzas National Monument, Florida, built 1742
- Fort Ticonderoga , New York, built 1759
- Hessian Powder Magazine, Carlisle, Pennsylvania, built 1777
- State Arsenal Complex, Charleston, South Carolina, built 1822
- Watervliet Arsenal, New York, built 1849
- Fort Wayne, Detroit, Michigan, built circa 1850
- Fort Sam Houston, San Antonio, Texas, built 1889

The powder magazine built in 1849 at the Watervliet Arsenal, New York, is pictured in Figures 87 through 89. The primary difference in construction was that the Watervliet building was stone on the exterior and brick on the interior.[80]

While some of the powder magazines in the US and the rest of the world have been restored, none match the immaculate condition of the 1857 powder magazine in the Benicia Arsenal.

Figure 86: Powder Magazine, Charleston SC, built 1712.
Source: Library of Congress, HABS, 1902 (left)
Halsey Map Preservation Society of Charleston (right).

Figure 87: Watervliet Powder Magazine Building 129, built 1849.
Source: Library of Congress, HAER NY-1E-1.

*Figure 88: Watervliet Powder Magazine door with date 1849.
Source: Library of Congress, HAER NY-1E-4.*

*Figure 89: Watervliet Powder Magazine interior, brick walls and ceiling.
Source: Library of Congress, HAER NY-1E-5.*

Chapter Sixteen

Magazine 10 Today

The Benicia Historical Museum has restored this building and installed a display of Benicia Arsenal artifacts. The interior is breathtaking from any view point (Figures 90 through 92). It has been used as a location for several retail catalogs and other photo opportunities. At the present time, the magazine is open to visitors by appointment only.

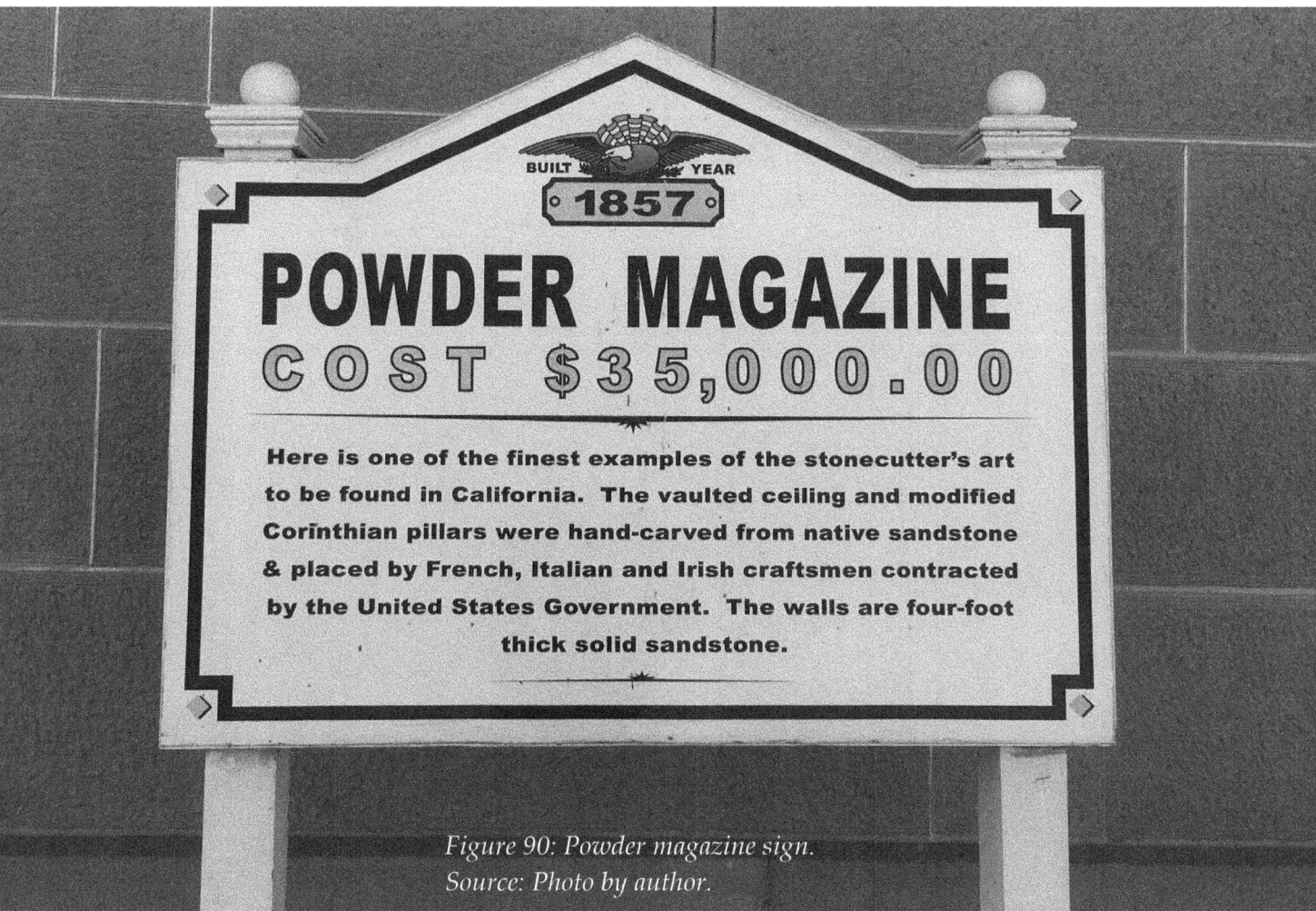

Figure 90: Powder magazine sign.
Source: Photo by author.

Figure 91: Powder Magazine 10 interior.
Source: Photo by author.

Figure 92: Grand ceiling view.
Source: Photo by author.

APPENDIX A

List of Men Who Built Powder Magazine 10

The ledger of Benicia Arsenal monthly activities from January 1857 to May 1858 provided a detailed account of the personnel that the Army hired, their duties, pay level, and days worked.

The worker's trade and pay level was as follows:

- Master mason: $6.00 per day
- Master stone cutter: $6.00 per day
- Stone cutter: $5.00 per day
- Slater (roof): $5.00 per day
- Mason/Stone dresser: $3.00 per day
- 1st class laborer: $2.50 per day
- 2nd class laborer: $2.25 per day

The following is a list of the 90 men who had a part in building Magazine 10 including birthplace and trade:

Name	Birthplace	Trade
Caleb S. Merrill	USA (MA)	Master mason-for building the magazine
Peter Larseneur	Canada	Master stone cutter
Charles Aened	?	2nd class laborer
Charles Arnold	France	2nd class laborer
Mathew Barry	Ireland	2nd class laborer
James Bashford	?	2nd class laborer
John Beckett	England	2nd class laborer
John Breslan	Ireland	Mason
John Buckley	Ireland	Mason
Frederick P. Burch	USA (CT)	Carpenter (roof)
Michael Burns	Ireland	2nd class laborer
Michael Cain	Ireland	1st class laborer/enlisted in Army
John Clarke	Wales	2nd class laborer
Joseph Cora	?	2nd class laborer
Aeneas Daly	Ireland	2nd class laborer
Joseph Damante	?	2nd class laborer
Robert Darling	Scotland	Stone cutter

Name	Origin	Occupation
Johan Derek	?	2nd class laborer
Joseph Diamond	USA (NY)	2nd class laborer
Thomas Dowling	Ireland	Stone cutter/enlisted in Army
D. D. Dunn	Ireland	2nd class laborer
Anthony Durkin	Ireland	2nd class laborer
John Farrell*	Ireland	2nd class laborer
John T. Farrell*	Ireland	Stone cutter
James Fitzgerald	Ireland	Mason/enlisted in Army
Lawrence O. Flattery	Ireland	2nd class laborer
Thomas Flynn	Ireland	Stone dresser
Patrick Gardiner	Ireland	1st class laborer and blacksmith
Peter German	Germany	Stone dresser
John C Gordon	Ireland	2nd class laborer/enlisted in Army
David Gorman	Ireland	Mason
John Hackett	Ireland	2nd class laborer
Jacob Hanson	?	2nd class laborer
Thomas Henny	Ireland	2nd class laborer
William Hogan	Ireland	2nd class laborer
John Hutchinson	England	2nd class laborer
Christopher Keenan	?	1st class laborer
James Keenan	Ireland	2nd class laborer
James Kelly	Ireland	Mason
John Kelley	Ireland	Mason
Peter King	England	2nd class laborer
James Langdon	England	Stone dresser
Charles Larseneur	Canada	Stone cutter
Louis Larseneur	Canada	Stone cutter
John Lofta	?	2nd class laborer
Sherbourne Locke	?	Mason
Michael Lorigan	Ireland	Mason
John Maloney	Ireland	2nd class laborer
James Manny	Ireland	Stone dresser
Robert McBride	Ireland	1st class laborer/re-enlisted in Army, Dec 1857
F. G McDonald	Ireland	2nd class laborer
William McDonald	Ireland	2nd class laborer/enlisted in Army
John McMullin	Ireland	2nd class laborer/enlisted in Army
Caleb S. Merrill Jr,	USA (MA)	Stone dresser
William G. Miller	?	2nd class laborer

Thomas Miller	?	2nd class laborer/enlisted in Army, Jan 1858
Thomas Mitchell	?	2nd class laborer
Thomas Moran	?	2nd class laborer/enlisted in Army
James Moreta	?	2nd class laborer
John Mullin	Ireland	2nd class laborer
James Murray	USA (NY)	2nd class laborer
Thomas Murray	Ireland	2nd class laborer
Andrew Nicholson	Scotland	2nd class laborer
John Norman	USA (NY)	2nd class laborer
John O'Conner	Ireland	Mason
Patrick Fitz Patrick	Ireland	2nd class laborer/enlisted in Army, Nov 1857
John Pattison	Sweden	2nd class laborer/enlisted with Army
Michael Pendergast	Ireland	2nd class laborer
John Practon	?	2nd class laborer
John Redmond	Ireland	1st class laborer
John H. Richardson	USA (NH)	Stone cutter
Robert Richardson	Scotland	Stone cutter
Larned Rico	?	2nd class laborer
John Sandro	USA (AL)	Stone cutter
Peter Scanlon	Ireland	Stone cutter
John Seery	Ireland	2nd class laborer
Charles Shannon	Ireland	2nd class laborer
Gustavers Shon	Germany	1st class laborer
James Sinclair	Scotland	Stone cutter
John Smiddy	Ireland	Stone dresser
James Smith	Ireland	2nd class laborer/enlisted in Army
Peter Larney Smith	Ireland	2nd class laborer/enlisted in Army
A. William Stannard	USA (CT)	2nd class laborer
John Sutherland	England	Mason
Frances Tracy	USA (MA)	Stone cutter
Henry Walsh	Ireland	2nd class laborer
James Walsh	Ireland	2nd class laborer
James Wells	?	2nd class laborer
Albion P. Whitman	USA (ME)	Mason/slater
Daniel Wynne	Ireland	Stone cutter

* Two men were named John Farrell, one was a stone cutter and one was a laborer (not related).

Appendix B

List of Inscriptions at Magazine 10

South Wall
A. B. H 1942
B Arsenal Cal
Dixon & Carla December 8th, 190?
(Guard) Lester V. Vanoni Concord, Calif May 19, 1955
J. F. Ashmore, C.R. Young June 26, 1916
Red Kinlaw Dunn, N. C. June 30
S. C. Fudge Hamlin, Texas April 23, 1943
Steve Millsap 10/15/18
William J Flannery July 9th 1878
WMG 1920 1942

North Wall
A. R. Ribeiro "1940"
A.H.B 1942
Alice O'C
Alvin L. Benas 9-2-19 S.F., Calif
AmBarty March 26, 1946
Antonio Cruz 06-30-00
Benicia Arsenal, Benicia, California
Bill Bennett Levittown, PA 5-16-61
Bill Cary
C Laught
C.E.B. 5-14-42
C.J.B. 1938
Cecil Dunwood Mar 13, 1944
Clifford Kirkpatrick 8-22-43
C. P. Miles Aug 9, 1955
David Wells 1960
Eleanor Ruiz July 1, 1944
Eulean Keeling
George W. Marshall July 29, 1914
Gerald Riley Martinez, Calif Oct 28 1941

H 1926
Harley Sullivan was last here 5-7-49
Howard Evans was here 1962
John Clarence June 16, 1939
Junior Garretty 9/10/40
KAI 1930
L.E.E. 7-2-42 Age 23
Leona Wieling 2126 E. 5th Street Benicia, Calif July 20, 1944
May Spanks Vallejo, CA July 20, 1944
Plummer Friston Knipas Dermarr
Richard W. Williams August 5, 1922
Richard Williams April 22, 1922
S.J.C. 1943
Saluting Igniting April 20, 1911
W R Mangles 1874
Walter Clarke May 12, 1942
Walter Kührt 28-III-1946
Wooden Bover Frankford Arsenal

East Wall
C.E. Young April 10, 1917 P.M.

East Columns
Red Kinlaw Dunn, NC 1930
Warren Philly June 30, 1917
Norman R MacGregor July 29, 1928
L. T. Hines Vallejo, Calif July 22 1958

Center Columns
A.B - 1944
H. Woodruff 3-6-61
J. Meadows 5-13-50
J.M. 1916
J.N.O. 5-10-43
Jack 12-29-19
Jack Warren 12/23/19
Jacob Jenson 10/10/17
JM 1901

KAT
L T Hines Vallejo July 22 1958
Norman MacGregor
Rafael Ventui 5-29-12
W. Sloane L-WY Project 8-15-61
W.J.O 1950

South-West Wall Column
C. P. Mills Aug 9, 1955
Josef J Dytza Benicia, Cal
M.D. 2-21-17 Jacob Jenson

Exterior East Wall
.X.X. Aug 1917 E & A 2nd Cal Inf
JM March 13, 1924
Bales 1932
April 1942 Tex Kid Windy

Works Cited

1. H. Dillon, 1980, *Great Expectations, The Story of Benicia, California*, p71
2. Robert Bruegmann, 1980, *Benicia: Portrait of an Early California Town, An Architectural History*, p73-74, 79
3. Ibid
4. Ibid
5. *Benicia Arsenal, Benicia, California*, Benicia Arsenal Public Information Officer. p24
6. Blaine Lamb, 2016, *The Extraordinary Life of Charles Stone*, Westholme Publishing, p43
7. *Benicia Arsenal, Benicia, California*, Benicia Arsenal Public Information Officer. p24
8. Cowell, J. W. 1963. *History of Benicia Arsenal: Benicia, California: January 1851 – December 1962*. Berkeley, Howell-North Books, p9-10
9. Ibid, p9
10. Ibid, p12-13
11. Eugene Bandel, 1932, *Frontier Life in the Army 1854-1861*, p 301
12. Cowell, J. W. 1963. *History of Benicia Arsenal: Benicia, California: January 1851 – December 1962*. Berkeley, Howell-North Books, p13
13. *Benicia Historical Museum Archives, Order 37,* Rules and Regulations for the Magazine, catalog 1999.099.0026
14. Cowell, J. W. 1963. *History of Benicia Arsenal: Benicia, California: January 1851 – December 1962*. Berkeley, Howell-North Books, p14
15. Blaine Lamb, 2016, *The Extraordinary Life of Charles Stone*, Westholme Publishing, p46
16. Cowell, J. W. 1963. *History of Benicia Arsenal: Benicia, California: January 1851 – December 1962*. Berkeley, Howell-North Books, p17
17. Blaine Lamb, 2016, *The Extraordinary Life of Charles Stone*, Westholme Publishing, p46
18. Ibid
19. Ibid
20. Ibid, p52
21. Ibid. p56
22. *The Charles P. Stone Journal*, Benicia Historical Museum, Catalog Number

2010.099.0033, 1856 Order No. 3

23 Catton, Bruce, *Mr. Lincoln's Army*, p70
24 Charles Pomeroy Stone, Wikipedia
25 Benicia Historical Museum Online Archives, *Powder Magazine Contract, October 30, 1855*, catalog 2019.021.0052
26 Samuel Merrill, Merrill Memorial, An Account of the Descendants of Nathaniel, an Early Settler of Newbury, Massachusetts, 1938, p451
27 *Boston Hide Droghers along the California Shores*, December 1929, Quarterly of the California Historical Society
28 *Black Gold in the Joaquin*, 1949, Frank F. Latta, Caxton Printers, p58-60
29 *Historical Encyclopedia of Illinois*, 1903, Warren County
30 *A Pioneer Gone, Caleb Merrill Jr. Obituary*, Hanford Journal Daily, Volume II, Number 24, 26 October 1898
31 Paul E. Vandor, *History of Fresno County, California With Biographical Sketches*, 1919, Volume 1 - p1251
32 Executive Documents, Second Session of the Thirty-third Congress, 1854-1855, Congressional Edition, Volume 788, *Contracts with the War Department*, page 42
33 H. Dillon, 1980, *Great Expectations, The Story of Benicia, California*, page 56-57
34 Family Tree: Severance-Merrill-James, 1634-1938, Merrill Papers, Bancroft Library.
35 *Daily Alta California*, San Francisco, 15 July 1865
36 *Daily Alta California*, San Francisco, 8 August 1866
37 *Black Gold in the Joaquin*, 1949, Frank F Latta, Caxton Printers, p58-60
38 U.S., Civil War Soldier Records and Profiles, 1861-1865, Asa C. Merrill
39 *Moffitt Book, Thumbnail sketches, deaths & obituaries*, Warren Co., Vol IV, p167
40 Caleb Strong Merrill Jr Obituary, *Hanford Journal Daily*, Vol II, No 24. 26 Oct 1898
41 Research Report Jacobs Engineering, April 1999, *Area M - Motor Pool and Historical Ordnance Storage Area*
42 Robert Bruegmann, 1980, *Benicia: Portrait of an Early California Town, An Architectural History*, p73-74
43 Research Report Jacobs Engineering, April 1999, *Area M - Motor Pool and Historical Ordnance Storage Area*
44 Christine Beall, 2004, *Masonry Design and Detailing*, McGraw-Hill
45 Great Valley Sequence, Wikipedia
46 Edward Salisbury Dana and William E. Ford, 1932, *Dana Textbook of Mineralogy*

47 Great Valley Sequence, Wikipedia
48 Richard Ralph D. Reed, 1951, *Geology of California*, p104
49 *Camel Barn Complex, Historic Structure Report*, July 1989, Architectural Resources Group, San Francisco, p13
50 Ibid, *Appendix, Petrological Analysis of the Camel Barn Complex*
51 W. A. Goodyear, 1890, *Report of the State Mineralogist for the Year Ending December 1, 1890*, California State Mining Bureau, Sacramento
52 Jasper Swann, March 2011, *Discovering Stone, Issue 19, Reproducing Historic Ashlar Finishes in Sandstone*, El1te Publishing Company, p24
53 *Camel Barn Complex, Historic Structure Report*, July 1989, Architectural Resources Group, San Francisco, p35
54 United States Federal Census, 1870, Benicia, Solano County, California
55 Louise Gomo obituary, September 25, 1908, Benicia New-era Herald
56 Register of Officers and Agents, Civil, Military and Naval in the Service of the United States, Benicia Arsenal, 30 September 1875, p6
57 Benicia Historical Museum, *John Gomo House*, catalog 2007.0046.0042
58 Arnold S. Lott, *A Long Line of Ships*, 1954, George Banta Publishing, p64
59 Sue Lemmon and Ernest D. Wichels, *Sidewheelers to Nuclear Power*, 1977, Leeward Publications, Annapolis, Maryland, p124
60 Passenger and Crew List, July 1845, Liverpool to New York
61 United States Federal Census, 1850, Tuolumne Township I, Tuolumne County, California
62 United States Federal Census, 1860, Vallejo Township, Solano County, California
63 Benicia Arsenal Ledger, June 1856 to June 1859, Benicia Historical Museum, catalog 1994.007.0002
64 Tennent, Thomas, 1890. *Tennent's Nautical Almanac*, San Francisco
65 Steve McKee, *Stone arches totally rock, Art and engineering meet in a little known Benicia building* by Benicia Herald, February 27, 2008
66 *Daily Alta California*, Volume 9, Number 127, 26 August 1857
67 Quoins, Wikipedia
68 Julia Bussinger and Beverly Phelan, 2004, *Benicia: Images of America*, p98
69 H. Dillon, 1980, *Great Expectations, The Story of Benicia, California*, page 71
70 Douglas E. Kyle, Hero Eugene Rensch, Ethel Grace Rensch, Mildred Brooke Hoover, William Abeloe (2002), *Historic Spots in California*, Fifth Edition, page 495
71 Research Report Jacobs Engineering, April 1999, *Area M - Motor Pool and Historical Ordnance Storage Area*, p 2M-3

72 Walter Kührt Personnel Archives, Federal Archives Unit PA 2.4, Berlin, Germany
73 The *San Francisco Examiner* August 13, 1922
74 Research Report Jacobs Engineering, April 1999, *Area M - Motor Pool and Historical Ordnance Storage Area*, p 2M-3
75 Benicia-Martinez Bridge System Project, Contra Costa and Solano Counties, Draft Environmental Impact Report, March 1995, p8
76 Ibid, p21-22
77 Vallejo Times Herald Online, September 10, 2017
78 Benicia-Martinez Bridge System Project, Contra Costa and Solano Counties, Final Environmental Impact Report, August 1997, Vol. II, Letter 49
79 Halsey Map Preservation Society of Charleston, Old Powder Magazine, p69
80 Robert Bruegmann, 1980, *Benicia: Portrait of an Early California Town, An Architectural History*, p74, 79

About the Author

Allan Gandy is a graduate of California Polytechnic State University, San Luis Obispo, with a degree in Metallurgical Engineering. He has been a resident of Benicia, California for over 40 years and has an interest in its history.

His first visit to Benicia was on a business trip in February 1975. He did not know about Benicia, but was impressed by the sign "Welcome to Historic Benicia." He and a colleague were there to inspect a valve at the Exxon Refinery. They had breakfast at Mable's Café on First Street and saw the city park and gazebo. They drove to the site and his colleague disassembled the valve, while his job as a metallurgist was to inspect the valve parts for signs of damage. One part needed some machining, so they found a shop nearby in the old Benicia Arsenal. As they drove down Park Road, they passed by the Camel Barns and Allan thought, "wow, those are beautiful."

A year and a half later, Allan moved to Benicia when changing jobs. He remembered the Camel Barns and drove for a visit. Behind them 200 yards, hidden amongst the low hills, was a rectangular stone building. There was a sculpture over the front entrance of an eagle on a cannon and the inscription "Erected 1857 AD." The building was secure so there was no way to see inside. Thirty years later he got to experience the beautiful interior of the powder magazine while on a guided tour of the museum.

In the author's own words, "This book has been a wonderful journey, taking me through different experiences as I was immersed in the lives of the powder magazine workers. After 160 years, the building stands strong with all its majestic beauty."

ABOOKS

ALIVE Book Publishing and ALIVE Publishing Group
are imprints of Advanced Publishing LLC,
3200 A Danville Blvd., Suite 204, Alamo, California 94507

Telephone: 925.837.7303
alivebookpublishing.com

www.ingramcontent.com/pod-product-compliance
Lightning Source LLC
Chambersburg PA
CBHW051548220426
43671CB00021B/2979